Vol. 1: H. Gross, **Quadratic Forms in Infinite-Dimensional Vector Spaces.** XXII, 419 pages, 1979
Vol. 2: F. Pham, **Singularités des systèmes différentiels de Gauss-Manin.** VI, 339 pages, 1979
Vol. 3: C. Okonek, M. Schneider, H. Spindler, **Vector Bundles on Complex Projective Spaces.** VIII, 389 pages, 1980
Vol. 4: **Complex Approximation,** Proceedings, Quebec, Canada, July, 3-8, 1978. Edited by Bernard Aupetit. X, 118 pages, 1980
Vol. 5: S. Helgason, **The Radon Transform.** X, 192 pages, 1980
Vol. 6: G. Lion, M. Vergne, **The Weil representation, Maslov index and Theta series.** VIII, 337 pages, 1980
Vol. 7: **Vector Bundles and Differential Equations,** Proceedings, Nice, France, June 12-17, 1979. Edited by André Hirschowitz. VI, 249 pages, 1980

Progress in Mathematics 7

Edited by
J. Coates and
S. Helgason

Vector Bundles and Differential Equations

Proceedings, Nice, France June 12-17, 1979

Edited by
André Hirschowitz

Birkhäuser
Boston, Basel, Stuttgart

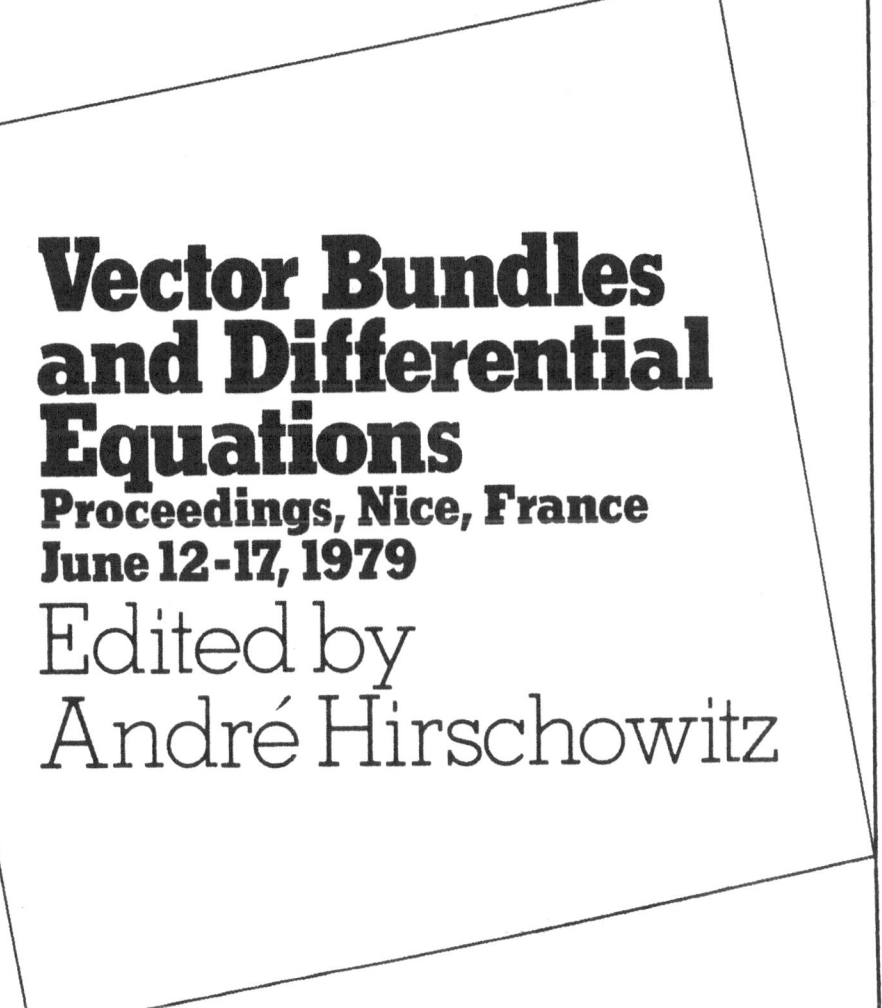

Editor

Professor André Hirschowitz
Université de Nice
Institut de Mathématiques
et Sciences physiques
Parc Valrose
06034 Nice Cedex
France

Library of Congress Cataloging in Publication Data
Main entry under title:

Vector bundles and differential equations.

(Progress in mathematics; 7)
Bibliography: p.
1. Vector bundles—Congresses. 2. Differential equations—Congresses.
I. Hirschowitz, A. II. Series: Progress in mathematics (Cambridge); 7.
QA612.63.V42 514'.224 80-19583

CIP—Kurztitelaufnahme der Deutschen Bibliothek

Vector bundles and differential equations:
proceedings, Nice, June 12-17, 1979 / ed. by A. Hirschowitz. —Boston, Basel, Stuttgart: Birkhäuser, 1980.
(Progress in mathematics; 7)

NE: Hirschowitz, André [Hrsg.]

ISBN 978-0-8176-3022-5 ISBN 978-1-4684-9415-0 (eBook)
DOI 10.1007/978-1-4684-9415-0

TABLE OF CONTENTS

P R E F A C E

This volume contains eight lectures resulting from papers delivered at the conference "Journées mathématiques sur les Fibrés vectoriels et Equations différentielles" held in Nice, France from June 12 through June 17, 1979.

The conference was sponsored by the Société Mathématique de France. Partial support was provided by:
Comité Doyen Jean Lépine de la Ville de Nice,
Conseil Général des Alpes Maritimes.

COUNTING SINGULARITIES OF QUADRATIC FORMS ON VECTOR BUNDLES

Wolf BARTH

0. INTRODUCTION.

The study of surfaces in $\mathbb{P}_3$ with many nodes (= ordinary double points) is a beautiful classical topic, which recently found much attention again [3, 4]. All systematic ways to produce such surfaces seem related to symmetric matrices of homogeneous polynomials or, more generally, to quadratic forms on vector bundles : If the form q on the bundle E is generic, then q is of maximal rank on an open set. The rank of q is one less on the discriminant hypersurface $\{\det q = o\}$, which represents the class $2c_1(E^*)$. This hypersurface is nonsingular in codimension one, but has ordinary double points in codimension two exactly where rank q drops one more step.

The aim of this paper is to show that the (rational homology class of the) singular variety of the discriminant is given by

(o) $$4(c_1c_2 - c_3), \qquad c_i = c_i(E^*).$$

If the base space has dimension three, the number of nodes of the discriminant surface is computed in this way.

Although I do not know of any place in the literature, where this formula can be found, I do not claim originality. If rank E = 2 for example, then $q \in \Gamma(S^2E^*)$, and the problem comes down to show that $c_3(S^2E^*) = 4c_1(E^*)c_2(E^*)$, which is well-known. Also, for morphisms $E \longrightarrow F$ there is Porteous' formula [9] expressing the loci of degeneration in terms of Chern classes. This formula does not apply directly to quadratic forms however, because they are selfadjoint, hence not generic as morphisms.

Formula (o) of course is some intersection number on the bundle space $\mathbb{P}(S^2E^*)$. Formulas for the higher-order degeneracies of quadratic forms analogous to Porteous' formula are to be expected as results of some computations in the intersection ring of $\mathbb{P}(S^2E^*)$.

I do not use here intersection theory on $\mathbb{P}(S^2E^*)$, partly because I had some trouble to identify the cycle on $\mathbb{P}(S^2E^*)$ of forms which on every fibre of E have a fixed given rank. My method is to associate with a quadratic form q a sheaf $\mathcal{C}$ on the discriminant hypersurface and to compute ch($\mathcal{C}$). This sheaf $\mathcal{C}$ is closely related to the theory of "even nodes" [4]

1. PRELIMINARIES.

Let S be the vector space of complex symmetric r x r matrices and define the following subvarieties

$$D := \{s \in S : \text{rank } s \leq r - 1\}$$

$$C := \{s \in S : \text{rank } s \leq r - 2\}$$

$$B := \{s \in S : \text{rank } s \leq r - 3\}$$

Then D is the zero-set of the determinant function, hence a hypersurface in S.

Lemma 1 : D _is nonsingular outside of_ C _and_ C _is nonsingular outside of_ B. _One has_

$$\text{codim}_S \; C = 3, \quad \text{codim}_S \; B = 6.$$

Proof : Put

$$s' = \begin{pmatrix} 0 & & & & \\ & 1 & & & \\ & & 1 & & \\ & & & \ddots & \\ & & & & 1 \end{pmatrix} \qquad s'' = \begin{pmatrix} 0 & & & & \\ & 0 & & & \\ & & 1 & & \\ & & & \ddots & \\ & & & & 1 \end{pmatrix}$$

Then each $s \in D\backslash C$ (resp. $s \in C\backslash B$) is of the form $as'a^{\dagger}$ (resp. $as''a^{\dagger}$) with $a \in GL(n)$. This shows that $D\backslash C$ (resp. $C\backslash B$) is homogeneous under GL(n), hence smooth. Also, the dimension of $C\backslash B$ is

$$n^2 - \dim \; \{a \in GL(n) : as''a^{\dagger} = s''\} .$$

Any $a \in GL(n)$ leaving s" invariant is of the form

$$a = \begin{pmatrix} a_1 & o \\ a_2 & a_3 \end{pmatrix} \qquad \begin{matrix} (a_1,a_2) \in \mathbb{C}^{2n} \\ a_3 \in O(n-2) \; . \end{matrix}$$

So the dimension to be subtracted is

$$2n + \dim O(n-2) = \frac{1}{2}\, n\,(n-1) + 3$$

and dim $C \setminus B = \frac{1}{2}\, n(n+1) - 3$. The same argument gives the dimension of B. ┘

Lemma 2 : D has ordinary quadratic singularities along C, i.e. any nonsingular (local) threefold meeting C transversally in a point $s_o \notin B$ intersects S in a surface with an ordinary doublepoint at s_o.

Proof : We may assume $s_o = s''$. Parametrize the threefold as $s(u_1,u_2,u_3) = (s_{ij}(u_1,u_2,u_3))$ with $s(o,o,o) = s''$. The intersection with S has the equation

$$f(u_1,u_2,u_3) = \det\,(s_{ij}(u_1,u_2,u_3)) = \sum_{i \neq j} s_{1i}\; s_{2j}\; s^{1i,2j},$$

where $s^{1i,2j}$ is the corresponding minor. So

$$\left.\frac{\partial^2 f}{\partial u_m\, \partial u_n}\right|_{o,o,o} = \frac{\partial s_{11}}{\partial u^m}\;\frac{\partial s_{22}}{\partial u^n} + \frac{\partial s_{11}}{\partial u^n}\;\frac{\partial s_{22}}{\partial u^m} - 2\;\frac{\partial s_{12}}{\partial u^m}\;\frac{\partial s_{12}}{\partial u^n}$$

and the hessian of f at (o,o,o) will be

$$(\partial s/\partial u)^t \begin{pmatrix} o & 1 & o \\ 1 & o & o \\ o & o & -2 \end{pmatrix} (\partial s/\partial u)$$

with

$$(\partial s/\partial u) := \begin{pmatrix} \partial s_{11}/\partial u_1 & \partial s_{11}/\partial u_2 & \partial s_{11}/\partial u_3 \\ \partial s_{22}/\partial u_1 & \partial s_{22}/\partial u_2 & \partial s_{22}/\partial u_3 \\ \partial s_{12}/\partial u_1 & \partial s_{12}/\partial u_2 & \partial s_{12}/\partial u_3 \end{pmatrix}$$

But the assumption that the threefold meets C transversally means rank $(\partial s/\partial u) = 3$. ┘

2. QUADRATIC FORMS ON VECTOR BUNDLES

A quadratic form on $\mathbb{C}^r$ can be thought of as a linear map $q : \mathbb{C}^r \longrightarrow (\mathbb{C}^r)^*$ with $q^\dagger = q$. This is the viewpoint for the study of quadratic forms on vector bundles to be used in the sequel.

So let X be a smooth projective threefold over $\mathbb{C}$ and E some rank-r vector bundle on X.

Definition : A quadratic form on E is linear morphism $q : E \longrightarrow E^*$ with $q^\dagger = q$. The set

$$\Delta := \{x \in E \ ; \ q(x) \text{ is not bijective}\}$$

is called the discriminant of q.

The quadratic forms on E form the vector space of sections in $S^2(E^*)$. If q is degenerate everywhere , Δ equals X, but in general Δ will be a surface. Δ can be empty only if $E = E^*$ and q is constant. The vector bundle $S^2(E^*)$ with typical fibre the space of symmetric $r \times r$ matrices contains as sub-fibre bundles the bundles

$$D(E), \ C(E), \text{ and } B(E),$$

the associated bundles with typical fibre D,C, and B. $s \in S^2(E^*)$ belongs to these subvarieties if rank $s \leq r-1$, $r-2$, and $r-3$ respectively.

For a quadratic form $q \in \Gamma(S^2E^*)$, Δ is the projection into X of $q \cap D(E)$.

Lemma 3 : (transversality) : Assume that q

- does not intersect $B(E)$,
- intersects $C(E)$ transversally (in finitely many points),
- intersects $D(E)$ transversally outside of $C(E)$.

Then Δ is a surface representing the class $2c_1(E^*)$. It is nonsingular except for finitely many nodes, the points x where rank $q(x) = r-2$.

Proof : a) Let $x_o \in X$ be a point with rank $q(x_o) = r-1$. By assumption q intersects $D(E)$ transversally near x_o, so $q \cap D(E)$ is nonsingular there. The projection $q \longrightarrow X$ being biregular, Δ will also be nonsingular near x_o. The equation $\det q = o$ vanishes to the first order on Δ in all but the finitely many point $x \in \Delta$ with rank $q(x) = r-2$. So the surface Δ represents in Pic X the class $\det E^* - \det E = 2c_1(E^*)$.

b) Let $x_o \in X$ be a point with rank $q(x_o) = r-2$. There is a neighborhood $U \subset X$ of x_o and a trivialization $E|U = U \times \mathbb{C}^r$ inducing trivializations

$$S^2(E^*)|U = U \times S,\ D(E)|U = U \times D,\ C(E)|U = U \times C.$$

Let $\pi = S \times U \longrightarrow S$ and $\rho = S \times U \longrightarrow U$ be the projections. The trivialization can be chosen such that $q(x_o) = s''$.

Now the equation for Δ near x_o is

$$\det(\pi\, q(x)) = 0.$$

By assumption, q intersects C(E) transversally at $q(x_o)$, so $\pi q(U)$ intersects C transversally at s'' and $\pi : q(U) \longrightarrow \pi q(U)$ is biregular near $q(x_o)$. Lemma 2 shows that $q(U) \cap D(E)$ is a surface with an ordinary node at $q(x_o)$. So Δ, the biregular image of this surface under ρ, will have an ordinary node at x_o. ┘

Lemma 4 (Bertini): Assume that $S^2(E^*)$ is generated by global sections. Then there is some Zariski-open subset of sections $q \in \Gamma(S^2(E^*))$ satisfying the conditions in Lemma 3.

Proof : Put $\Gamma := \Gamma(S^2(E^*))$ and consider the evaluation map $\gamma : X \times \Gamma \longrightarrow S^2(E^*)$. This map γ is regular everywhere, so the subvarieties

$$\tilde{D} = \gamma^{-1}D(E),\quad \tilde{C} = \gamma^{-1}C(E),\quad \tilde{B} = \gamma^{-1}B(E)$$

of $X \times \Gamma$ have codimension 1,3 and 6 respectively. Denote by $\pi : X \times \Gamma \longrightarrow \Gamma$ the (proper) projection and define subvarieties of Γ as follows :

i) $\Gamma_1 := \pi(\tilde{B})$. This is a subvariety of Γ with codimension ≥ 3, because $\dim X = 3$.

ii) Let $C' \subset \tilde{C}$ be the subvariety of points where $d(\pi|\tilde{C})$ is not surjective. Then $\Gamma_2 := \pi(C')$ is a subvariety of codimension ≥ 1.

iii) Let $D' \subset \tilde{D}\setminus\tilde{C}$ be the subvariety of points where $d(\pi|\tilde{D}\setminus\tilde{C})$ is not surjective and let $\bar{D}$ be its closure in $X \times \Gamma$. Then $\Gamma_3 := \pi(\bar{D})$ again is a subvariety of Γ of codimension ≥ 1.

Now the Zariski-open subset can be taken as the complement of $\Gamma_1 \cup \Gamma_2 \cup \Gamma_3$. ┘

Combining lemmas 3 and 4 one obtains :

Proposition 1 : _If_ $S^2(E^*)$ _is spanned by global sections, then for general_ $q \in \Gamma(S^2(E^*))$ _the discriminant_ $\Delta \subset X$ _is a nonsingular surface except for finitely many nodes. It represents the class_ $2c_1(E^*)$.

3. THE COKERNEL OF q

Now let $q = E \longrightarrow E^*$ be a quadratic form which is general in the sense of proposition 1. Outside of Δ , the morphism q is an isomorphism.

So there is an exact sequence

$$o \longrightarrow E \xrightarrow{q} E^* \longrightarrow \mathcal{C} \longrightarrow o$$

with an $\mathcal{O}_X$-sheaf $\mathcal{C}$ supported on Δ . Next we shall analyze the cokernel $\mathcal{C}$.

Denote by $\{x_i\}$ the finite set of nodes of Δ .

Lemma 5 : _Outside of_ $\{x_i\}$, _the sheaf_ $\mathcal{C}$ _is an invertible_ $\mathcal{O}_\Delta$_-sheaf_.

Proof : Fix some point $x_o \in \Delta \setminus \{x_i\}$ and let f be a local equation for Δ near x_o. Since rank $(q|\Delta)$ = r-1 near x_o, there is a section e_1 in E, without zeroes, such that $q(e_1)|\Delta = o$. This section e_1 can be extented to a basis $e_1,\ldots,e_r$ for E near x_o. In the basis for E and an arbitrary one for E^* write

$$q = \begin{pmatrix} q_{11} & \cdots & q_{1r} \\ \vdots & & \vdots \\ q_{r1} & & q_{rr} \end{pmatrix}$$

then $q(e_1)$ is the vector $(q_{11},\ldots,q_{r1})$. Since it vanishes on Δ , we can write $q_{j1} = f.q'_{j1}$ and $q = q' \circ \varphi$ with

$$q' = \begin{pmatrix} q'_{11} & \cdots & q_{1r} \\ \vdots & & \vdots \\ q'_{r1} & \cdots & q_{rr} \end{pmatrix} \qquad \varphi = \begin{pmatrix} f & & & o \\ & 1 & & \\ & & \ddots & \\ o & & & 1 \end{pmatrix}$$

Since $\det q = f \det q'$ vanishes on Δ to the first order only, $\det q'$ cannot vanish near x_o and q' is an isomorphism of E onto E^*.

The diagram

$$\begin{array}{ccccccccc} o & \longrightarrow & E & \xrightarrow{\varphi} & E & \longrightarrow & \mathcal{O}_\Delta & \longrightarrow & o \\ & & \| & & \downarrow q' & & & & \\ o & \longrightarrow & E & \xrightarrow{q} & E^* & \longrightarrow & \mathcal{C} & \longrightarrow & o \end{array}$$

then shows that $\mathcal{C} = \mathcal{O}_\Delta$ near x_o.

To understand the situation near the singularities x_i, we shall blow them up : Fix some x_i and let $\sigma_i : \tilde{X}_i \longrightarrow X$ be the monoidal transform with center x_i. The surface $\Sigma_i := \sigma_i^{-1} x_i$ then is a copy of $\mathbb{P}_2$ with self-intersection $\Sigma_i^2 = -h_i$, h_i the positive generator of $H^2(\mathbb{P}_2,\mathbb{Z})$. Since x_i was an ordinary node of Δ, the proper transform $\tilde{\Delta}_i \subset \tilde{X}_i$ of Δ is nonsingular near Σ_i. It intersects Σ_i in a curve C_i, which is a non-degenerate conic on Σ_i and has on $\tilde{\Delta}_i$ self-intersection -2.

Additionally, it is no loss of generality to assume $r = 2$ (locally near x_i). In fact, there is a basis for E_{x_i} such that $q(x_i)$ looks like s". There is a rank-(r-2) subbundle $G \subset E$ near x_i restricting in x_i to the subspace of E_{x_i} spanned by the last r-2 basis vectors. So $q|G$ is non-degenerate near x_i. Define $F \subset E$ as the subbundle $G^\perp$, i.e. the kernel of $E \xrightarrow{q} E^* \longrightarrow G^*$. Then locally near x_i, E is an orthogonal direct sum $F \oplus G$ and

$$q = \begin{pmatrix} q_F & o \\ o & q_G \end{pmatrix}$$

with respect to this decomposition. q_G being an isomorphism, the original cokernel $\mathcal{C}$ is isomorphic to the cokernel of $q_F : F \longrightarrow F^*$, with rank F = 2.

So replace E by $F = 2\mathcal{O}$ and write

$$q = \begin{pmatrix} a & c \\ c & b \end{pmatrix}$$

with functions a,b,c vanishing at x_i. Let $\tilde{a},\tilde{b},\tilde{c},\tilde{q}$ be the pullbacks of a,b,c,q to X_i. On $\tilde{X}_i$ there is near Σ_i a diagram of exact sequences

$$\begin{array}{ccccccccc}
 & & & & 0 & & 0 & & \\
 & & & & \downarrow & & \downarrow & & \\
0 & \longrightarrow & 2\mathcal{O}_{\tilde{X}_i} & \xrightarrow{p} & 2\mathcal{I}_{\Sigma_i} & \longrightarrow & M_i & \longrightarrow & 0 \\
 & & \| & & \downarrow & & \downarrow & & \\
0 & \longrightarrow & 2\mathcal{O}_{\tilde{X}_i} & \xrightarrow{\tilde{q}} & 2\mathcal{O}_{\tilde{X}_i} & \longrightarrow & \tilde{\mathcal{C}}_i & \longrightarrow & 0 \\
 & & & & \downarrow & & \downarrow & & \\
 & & & & 2\mathcal{O}_{\Sigma_i} & = & 2\mathcal{O}_{\Sigma_i} & & \\
 & & & & \downarrow & & \downarrow & & \\
 & & & & 0 & & 0 & &
\end{array}$$

Now $2\mathcal{I}_{\Sigma_i}$ is locally free, and p, the map induced by $\tilde{q}$, is given by a matrix

$$\begin{pmatrix} a/g & c/g \\ c/g & b/g \end{pmatrix} ,$$

g a local equation for Σ_i. Also det $\tilde{q}$ vanishes on Σ_i outside of C_i only to order 2. So det p does not vanish there at all. Since det p vanishes only along $\tilde{\Delta}_i$, and there of order one, a modification of lemma 5 shows that M_i is an invertible $\mathcal{O}_{\tilde{\Delta}_i}$-sheaf. Outside of C_i, the morphism $M_i \longrightarrow \mathcal{C}_i = \sigma^*\mathcal{C}$ is an isomorphism.

To formulate the result, let $\sigma : \tilde{X} \to X$ be the simultaneous blow up of all x_i, let $\tilde{q} : \tilde{E} \to \tilde{E}^*$ be the pullback of q, let $\Sigma = \cup\, \Sigma_i$ be the union of the exceptional planes, $\tilde{\Delta}$ the proper transform of Δ, and $C = \cup\, C_i = \Sigma \cap \tilde{\Delta}$ the union of the conics.

Proposition 2 : *On $\tilde{X}$ the quadratic form q induces an exact sequence*

(1) $$0 \longrightarrow \tilde{E} \xrightarrow{\tilde{q}} \tilde{E}^* \longrightarrow \tilde{\mathcal{C}} \longrightarrow 0 ,$$

and the cokernel $\tilde{\mathcal{C}}$ is an extension

(2) $$0 \longrightarrow M \longrightarrow \tilde{\mathcal{C}} \longrightarrow 2\mathcal{O}_{\Sigma} \longrightarrow 0$$

with an invertible $\mathcal{O}_{\tilde{\Delta}}$-sheaf M.

Proof : Outside of Σ , M is the sheaf $q^*\mathcal{C}$. Over each conic C_i, it extends by the sheaf M_i constructed above. ┘

The next proposition is a reformulation of the symmetry of q.

Proposition 3 : <u>There is an isomorphism</u>

$$\varepsilon : \mathcal{C} \to \mathcal{E}xt^1_{\mathcal{O}_{\tilde{X}}}(\tilde{\mathcal{C}}, \mathcal{O}_{\tilde{X}})$$

<u>inducing an isomorphism</u>

(3) $$M^{\otimes 2} = \mathcal{O}_{\tilde{\Delta}}(\tilde{\Delta} - \Sigma).$$

Proof : By virtue of $\tilde{q} = \tilde{q}^{t}$, there is a commutative diagram

$$\begin{array}{ccccccccc} 0 & \longrightarrow & \tilde{E} & \xrightarrow{\tilde{q}} & \tilde{E}^* & \longrightarrow & \tilde{\mathcal{C}} & \longrightarrow & 0 \\ & & \| & & \| & & & & \\ 0 & \longrightarrow & \tilde{E} & \xrightarrow{\tilde{q}^t} & \tilde{E}^* & \longrightarrow & \mathcal{E}xt^1_{\mathcal{O}_{\tilde{X}}}(\tilde{\mathcal{C}}, \mathcal{O}_{\tilde{X}}) & \longrightarrow & 0 \end{array}$$

inducing the isomorphism ε .

Now the dual sequence of (2) is

$$\begin{array}{ccccccccc} 0 & \longrightarrow & \mathcal{E}xt^1_{\mathcal{O}_{\tilde{X}}}(2\mathcal{O}_\Sigma, \mathcal{O}_{\tilde{X}}) & \longrightarrow & \mathcal{E}xt^1_{\mathcal{O}_{\tilde{X}}}(\tilde{\mathcal{C}}, \mathcal{O}_{\tilde{X}}) & \longrightarrow & \mathcal{E}xt^1_{\mathcal{O}_{\tilde{X}}}(M, \mathcal{O}_{\tilde{X}}) & \longrightarrow & 0 \\ & & \| & & \| & & \| & & \\ 0 & \longrightarrow & 2\mathcal{O}_\Sigma(\Sigma) & \longrightarrow & \tilde{\mathcal{C}} & \longrightarrow & M^* \otimes \mathcal{O}_{\tilde{\Delta}}(\tilde{\Delta}) & \longrightarrow & 0 \end{array}$$

Since the map

$$2\mathcal{O}_\Sigma(\Sigma) \longrightarrow \tilde{\mathcal{C}} \longrightarrow 2\mathcal{O}_\Sigma$$

is injective on Σ outside of C, it induces an exact sequence

$$0 \longrightarrow 2\mathcal{O}_\Sigma(\Sigma) \longrightarrow 2\mathcal{O}_\Sigma \longrightarrow \mathcal{O}_C \longrightarrow 0$$

and a diagram

$$
\begin{array}{ccccccccc}
 & & & & 0 & & 0 & & \\
 & & & & \downarrow & & \downarrow & & \\
 & & & & M & = & M & & \\
 & & & & \downarrow & & \downarrow & & \\
0 & \longrightarrow & 2\mathcal{O}_{\Sigma}(\Sigma) & \longrightarrow & \tilde{\mathcal{E}} & \longrightarrow & M^{*}\otimes\mathcal{O}_{\tilde{\Delta}}(\tilde{\Delta}) & \longrightarrow & 0 \\
 & & \| & & \downarrow & & \downarrow & & \\
0 & \longrightarrow & 2\mathcal{O}_{\Sigma}(\Sigma) & \longrightarrow & 2\mathcal{O}_{\Sigma} & \longrightarrow & \mathcal{O}_{C} & \longrightarrow & 0 \\
 & & & & \downarrow & & \downarrow & & \\
 & & & & 0 & & 0 & &
\end{array}
$$

The righthand column implies

$$M^{*} \otimes \mathcal{O}_{\tilde{\Delta}}(\tilde{\Delta}) = M \otimes \mathcal{O}_{\tilde{\Delta}}(C). \qquad \lrcorner$$

Corollary : <u>For</u> $N := M \otimes \sigma^{*}(\det E) \otimes \mathcal{O}_{\tilde{\Delta}}(2C)$ <u>one has</u>

$$N^{\otimes 2} = \mathcal{O}_{\tilde{\Delta}}(C),$$

<u>i.e., the divisor class</u> C <u>on</u> $\tilde{\Delta}$ <u>is divisible by</u> 2.

Proof : $\mathcal{O}_{\tilde{\Delta}}(\tilde{\Delta}) = \sigma^{*}(2c_1 E^{*}) \otimes \mathcal{O}_{\tilde{\Delta}}(-2C)$ and $M^{\otimes 2} = \sigma^{*}(\det E^{*})^{\otimes 2} \otimes \mathcal{O}_{\tilde{\Delta}}(-3C)$. $\lrcorner$

4. APPLICATION OF GROTHENDIECK-RIEMANN-ROCH.

This section gives the number of singularities x_i in terms of the Chern-classes $c_i = c_i(E^{*})$. In fact, all one has to do is to compute Σ^3, because of

$$\Sigma^3 = \sum_i \Sigma_i^3$$

and $\Sigma_i^3 = 1$ for all i. (We shall denote by c_i also the pull-back $\sigma^{*} c_i \in H^2(\tilde{X},\mathbb{Z})$). Consider the two exact sequences (1) and (2). The additivity of the Chern-character [7, Appendix A] shows.

$$\text{(4)} \qquad ch(E^{*}) - ch(E) = ch(\tilde{\mathcal{E}}) = ch(M) + ch(2\mathcal{O}_{\Sigma}).$$

E^{*} being locally free, one has the well-known formula [7, p. 432]

$$ch(E^{*}) = r + c_1 + \frac{1}{2}(c_1^2 - c_2) + \frac{1}{6}(c_1^3 - 3c_1c_2 + 3c_3).$$

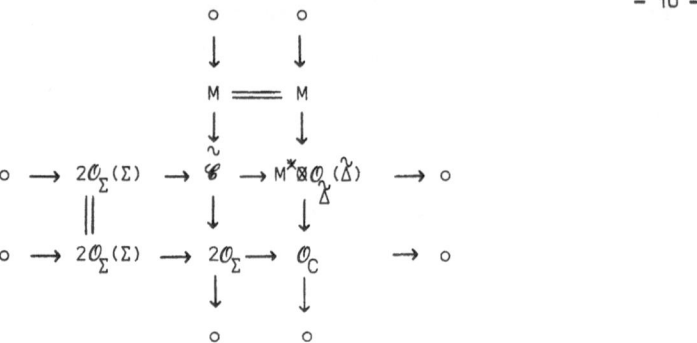

Because of $c_i(E) = (-1)^i \; c_i(E^*)$ we find for the codimension -3 component of $\widetilde{ch}(\mathcal{C})$

(5) $\quad \widetilde{ch}_3(\mathcal{C}) = \frac{1}{3} c_1^3 - c_1 c_2 + c_3.$

The computation of $ch(2\mathcal{O}_\Sigma) = 2\, ch(\mathcal{O}_\Sigma)$ is easy too : Because of the exact sequence

$$0 \longrightarrow \mathcal{O}_{\tilde{X}}(-\Sigma) \longrightarrow \mathcal{O}_{\tilde{X}} \longrightarrow \mathcal{O}_\Sigma \longrightarrow 0$$

one has

$$ch(\mathcal{O}_\Sigma) = ch(\mathcal{O}_{\tilde{X}}) - ch(\mathcal{O}_{\tilde{X}}(-\Sigma)) = 1 - (1 - \Sigma + \frac{1}{2}\Sigma^2 - \frac{1}{6}\Sigma^3)$$

and

(6) $\quad ch_3(2\mathcal{O}_\Sigma) = \frac{1}{3} \; \Sigma^3.$

The complicated part is to compute $ch(M)$ by $\mathcal{G}$r RR for the embedding $i : \tilde{\Delta} \longrightarrow \tilde{X}$,

$$ch(M) = i_* \left[ch(M|\tilde{\Delta}).\; td(-N_{\tilde{\Delta}/\tilde{X}})\right].$$

Here (intersections are taken in $H^*(\tilde{\Delta},\mathbb{Q})$)

$$\begin{aligned}
ch(M|\tilde{\Delta}) &= 1 + M + \frac{1}{2} M^2 \\
c(N_{\tilde{\Delta}/\tilde{X}}) &= 1 + \mathcal{O}_{\tilde{\Delta}}(\tilde{\Delta}) \\
c(-N_{\tilde{\Delta}/\tilde{X}}) &= 1 - \mathcal{O}_{\tilde{\Delta}}(\tilde{\Delta}) + \mathcal{O}_{\tilde{\Delta}}(\tilde{\Delta})^2 \\
td(-N_{\tilde{\Delta}/\tilde{X}}) &= 1 + \frac{1}{2} c_1(-N_{\tilde{\Delta}/\tilde{X}}) + \frac{1}{12}(c_1^2 + c_2)(-N_{\tilde{\Delta}/\tilde{X}}) = \\
&= 1 - \frac{1}{2}\mathcal{O}_{\tilde{\Delta}}(\tilde{\Delta}) + \frac{1}{6}\mathcal{O}_{\tilde{\Delta}}(\tilde{\Delta})^2 \\
ch_3(M) &= i_*\left[ch(M|\tilde{\Delta}) \;.\; td(-N_{\tilde{\Delta}/\tilde{X}})\right]_2 = \\
&= \frac{1}{6}\tilde{\Delta}^3 - \frac{1}{2}(M.\mathcal{O}_{\tilde{\Delta}}(\tilde{\Delta})) + \frac{1}{2}(M.M).
\end{aligned}$$

Now using $\mathcal{O}_{\tilde{X}}(\tilde{\Delta}) = 2c_1 - 2\mathcal{O}_{\tilde{X}}(\Sigma)$ and

$$M = \frac{1}{2}(2c_1 - 3\mathcal{O}_{\tilde{\Delta}}(\Sigma)) = c_1 - \frac{3}{2}\mathcal{O}_{\tilde{\Delta}}(\Sigma)$$

by formula (3), we compute

$$\tilde{\Delta}^3 = 8\, c_1^3 - 8\ \Sigma^3$$

$$(M.\mathcal{O}_{\tilde{\Delta}}(\tilde{\Delta})) = 4\, c_1^3 - 6\, \Sigma^3$$

$$(M.M) = 2\, c_1^3 - \frac{9}{2}\, \Sigma^3$$

(7) $$ch_3(M) = \frac{1}{3}\, c_1^3 - \frac{7}{12}\, \Sigma^3 .$$

Then substituting (5), (6), and (7) in formula (4), one obtains finally

$$\frac{1}{3}\ c_1^3 - c_1c_2 + c_3 = \frac{1}{3}\, c_1^3 - \frac{7}{12}\, \Sigma^3 + \frac{1}{3}\ \Sigma^3,$$

or

$$\Sigma^3 = 4(c_1c_2 - c_3) \quad .$$

Theorem : <u>In the situation of proposition 1, the number of nodes of</u> Δ <u>equals</u> $4(c_1c_2 - c_3)$. <u>Here</u> $c_i = c_i(E^*)$, $i = 1,2,3$.

Remark : If rank $E = 2$, then $c_3 = o$ and the formula gives $4\, c_1c_2$ for the number of nodes. In fact, in this case the nodes are exactly the points where $q \in \Gamma(S^2E^*)$ vanishes and $c_3(S^2E^*) = 4c_1(E^*)c_2(E^*)$ is well-known.

5. GENERALISATIONS.

Next two possible generalisations of the formula in the theorem above are given without proofs.

a) Of course, it is not necessary to assume that $\dim X = 3$. Also if $\dim X > 3$, the discriminant hypersurface of a sufficiently general quadratic form $q : E \longrightarrow E^*$ is nonsingular in codimension 1, it represents the class $2c_1(E^*)$, and its singularities form a cycle representing the class $4(c_1(E^*)c_2(E^*)-c_3(E^*))$.

b) Very often one meets twisted quadratic forms : A <u>quadratic form</u> on E <u>with</u> values in L, some line bundle on X, is a linear morphism

$$q = E \longrightarrow E^* \otimes L,$$

which is symmetric in the sense that

$$q = q^{\dagger} \otimes id_L .$$

The same machinery applies : If q is sufficiently general, then the discriminant Δ is a hypersurface representing the class $2c_1(E^*) + r\lambda$, where $\lambda = c_1(L)$ and $r = \text{rank } E$.

Δ is nonsingular in codimension 1, with only finitely many nodes (if dim X = 3). On Δ we have the line bundle M again, but (3) now changes into

$$M^{\otimes 2} = \mathcal{O}_{\tilde\Delta}(\tilde\Delta - \Sigma) \otimes L.$$

Putting $N := M \otimes \sigma^*(\det E) \otimes \mathcal{O}_{\tilde\Delta}(2C)$ again, we have

$$N^{\otimes 2} = \mathcal{O}_{\tilde\Delta}(C) \otimes L^{\otimes(r+1)}. \tag{8}$$

So the divisor class of C on $\tilde\Delta$ is a square if and only if $L^{\otimes(r+1)}|C$ is one, e.g. if r is odd.

Definition [4,Def.2.4] : The set of nodes of Δ is called <u>strictly even</u> (<u>weakly even</u>) if C is a square (if there is a line bundle H on X such that $H \otimes \mathcal{O}_{\tilde X}(C)$ is a square).

The simple formula $4(c_1c_2 - c_3)$ for the number of nodes of Δ , which does not even involve r, change drastically. It becomes

$$4(c_1c_2 - c_3) + 4c_2.\lambda + 2(r-1)c_1^2.\lambda + \\ + (r^2-1)c_1\,\lambda^2 + \binom{r+1}{3}\lambda^3, \tag{9}$$

where

$$c_i = c_i(E^*),\ \lambda = c_1(L).$$

6. APPLICATION TO THE VARIETY OF JUMPING LINES OF A VECTOR BUNDLE ON $\mathbb{P}_3$.

Let F on $\mathbb{P}_3$ be an instanton bundle. This means

$$\text{rank } F = 2,$$

$$c_1(F) = o,$$

$$h^o(F) = o,$$

$$h^1(F(-2)) = o.$$

Then it is known that F is the cohomology of a monad

$$n\,\mathcal{O}_{\mathbb{P}}(-1) \xrightarrow{a} (2n+2)\,\mathcal{O}_{\mathbb{P}} \xrightarrow{a^t} n\,\mathcal{O}_{\mathbb{P}}(1)$$

with $n = c_2(F)$. Here

$$a = (a_{ij}(z)),\ i = 1,\dots,2n+2,\ j = 1,\dots,n,$$
$$a_{ij}(z) \in \Gamma(\mathcal{O}_{\mathbb{P}_3}(1)).$$

Now the condition $a^\dagger \circ a = o$ is equivalent to

$$a^\dagger(z_1)a(z_2) + a^\dagger(z_2)a(z_1) = o$$

for all z_1, z_2. So $q(z_1,z_2) := a^\dagger(z_1)a(z_2)$ is alternating and bilinear in z_1, z_2. We can think of it as depending linearly on $z_1 \wedge z_2$, say $Q(z_1 \wedge z_2)$. Now $a^\dagger$ is formed with respect to a <u>symplectic</u> form on $(2n+2)\mathcal{O}_{\mathbb{P}_3}$. This implies $(a^\dagger)^\dagger = -a$ and therefore $Q^\dagger(z_1 \wedge z_2) = Q(z_1 \wedge z_2)$. If $\mathbb{P}_5$ denotes the projective space, in which Grass (1;3) is embedded by its Plücker embedding, then Q is nothing but a twisted quadratic form

$$Q: n\,\mathcal{O}_{\mathbb{P}_5}(-1) \longrightarrow n\,\mathcal{O}_{\mathbb{P}_5}\ .$$

In fact (without proof) the cokernel of $Q|$Grass is isomorphic with the direct image sheaf $q_{*1}\,p^*F(-1)$, where $q: F(1,2,3) \longrightarrow$ Grass and $p: F(1,2,3) \longrightarrow \mathbb{P}_3$ are the standard projections (cf. [1, section 4]). Also, if $R \subset \mathbb{P}_3$ is a line corresponding to $z_1 \wedge z_2$, then

$$\text{rank } Q(z_1 \wedge z_2) = n-k$$

if and only if

$$F|R = \mathcal{O}_R(k) \oplus \mathcal{O}_R(-k).$$

So the discriminant $\Delta \subset$ Grass of $Q|$Grass is nothing but the variety of jumping lines of F. And if Q should happen to be "sufficiently general" then our machinery developed so far predicts that the threefold Δ is nonsingular in codimension one, but has a curve C of double points corresponding to lines R with

$$F|R : \mathcal{O}_R(2) \oplus \mathcal{O}_R(-2).$$

These are the higher order jumping lines.

Formula (9) can be applied to this situation. Just tensor Q by $\mathcal{O}_{\mathbb{P}_5}(1)$ and put

$$E = E^* = n\ \mathcal{O}_{\mathbb{P}},\ L = \mathcal{O}_{\mathbb{P}}(1).$$

Then it predicts that C is the intersection of Grass with a surface $C' \subset \mathbb{P}_5$ of degree $\binom{n+1}{3}$. The first three cases are

$$\begin{array}{lllll} n & = 1, & 2, & 3 & \\ \deg\ C' & = o, & 1, & 4 & . \end{array}$$

For n = 1,2 this matches perfectly with experience [8, prop. 9.11] , but I do not know the degree of C for a single instanton bundle with n = 3 even.

I should mention my original hope that the degree of C should be an invariant distinguishing the different components of stable rank-2 bundles F on $\mathbb{P}_3$ with the same Chern classes $c_1 = o$, $c_2 = n$ (not only instanton bundles). But in general, deg C will depend only on $ch(q_{*1}p^*F(-1))$, which in its turn is determined by the Chern classes of F via the relative Riemann Roch.

7. SURFACES IN $\mathbb{P}_3$ WITH MANY NODES.

This section contains some examples of discriminant surfaces in $\mathbb{P}_3$.

a) Symmetric determinants of linear forms.

Let $a_{ij}(z)$, i, j=1,...,n, be linear forms in z_o, z_1, z_2, z_3, where $a_{ij} = a_{ji}$. The surface Δ of degree n given by the vanishing of $\det(a_{ij}(z))$ is the discriminant of the morphism

$$(a_{ij}(z)) = n\,\mathcal{O}_{\mathbb{P}_3} \longrightarrow n\,\mathcal{O}_{\mathbb{P}_3}(1).$$

For general $(a_{ij}(z))$ our machinery applies and (9) shows that the number of nodes on Δ equals $\binom{n+1}{3}$. By formula (8), the set of nodes is even if n is odd, and at least weakly even if n is even. (These facts are well-known, see [4] for modern proofs).

For small n one has this table

n	number of nodes	surface
2	1	nondegenerate quadratic cone
3	4	quadrinodal cubic of Cayley
4	10	symmetroid
5	20	Quintics of Gallarati [5] and Beauville
6	35	

b) Quartics

Except for sums of line bundles, the only bundles I know on $\mathbb{P}_3$ with $c_1 \leq 2$, and generated by global sections, are the following ones :

bundle	rank	c_1	c_2	c_3
$T_{\mathbb{P}}(-1)$	3	1	1	1
$2T_{\mathbb{P}}(-1)$	6	2	3	4
$\Omega_{\mathbb{P}}(2)$	3	2	2	o
$V_o(1) = \Omega_{\mathbb{P}}(2)/\mathcal{O}_{\mathbb{P}}$	2	2	2	o
F_1	3	2	2	2
F_2	3	2	3	4
F_3	3	2	4	8

Glossary : V_o is the null-correlation bundle [1] , F_1, F_2 and F_3 are obtained by Vogelaar's construction [11] as extensions

$$o \longrightarrow 2\mathcal{O}_{\mathbb{P}} \longrightarrow F_i \longrightarrow \mathcal{I}_{C_i}(2) \longrightarrow o,$$

where

C_1 = nondegenerate conic,

C_2 = cubic rational norm curve,

C_3 = elliptic quartic curve.

Using all these bundles, one can produce only quartics with 6,8,10 or 16 nodes. The surface with 10 nodes is the symmetroid from a). The surface with 16 nodes (Kummer's surface) can be obtained from $E^* = \Omega(2)$ or $V_o(1)$.

A quadratic form

$$\mathcal{O}(-1) + \mathcal{O} \longrightarrow \mathcal{O}(1) + \mathcal{O}(2)$$

gives 6 nodes and a form

$$\mathcal{O}(-1) + \mathcal{O}(-1) \longrightarrow \mathcal{O}(1) + \mathcal{O}(1)$$

gives 8 nodes.

c) Quintics

Beauville [3] showed that the only possible numbers for even sets of nodes on otherwise nonsingular quintics are 16 or 20. The quintics from a) above have 20 even nodes. Those with 16 even nodes are obtained as discriminants of

$$q: \mathcal{O}_{\mathbb{P}}(-1) + 2\mathcal{O}_{\mathbb{P}} \longrightarrow 2\mathcal{O}_{\mathbb{P}}(1) + \mathcal{O}_{\mathbb{P}}(2)$$

These quintics were discovered first by Togliatti [10] projecting a cubic hypersurface in $\mathbb{P}_5$ onto $\mathbb{P}_3$ out of a line contained in the hypersurface. The most interesting ones however are those acquiring 15 more nodes, which come from points x with rank $q(x) = 2$, but where q touches D(E).

d) Sextics

There is some reason [2] to search for sextics with 48 even nodes. But it seems impossible to find such surfaces as discriminants of sufficiently general quadratic forms. The problem one always meets when trying this, is that bundles with small enough Chern classes usually are not globally generated.

It is known however [6,p.53] that there are sextics with up to 40 even nodes. Let me conclude by collecting some example

r	E^*	c_1	c_2	c_3	λ	number of nodes	even ?
2	$\mathcal{O}+\mathcal{O}(2)$	2	o	o	1	15	weakly
2	$\mathcal{O}(1) + \mathcal{O}(2)$	3	2	2	o	24	strictly
2	$2\mathcal{O}(1)$	2	1	o	1	27	weakly
4	$3\mathcal{O} + \mathcal{O}(1)$	1	o	o	1	31	weakly
3	$3\mathcal{O}(1)$	3	3	1	o	32	strictly
6	$6\mathcal{O}$	o	o	o	1	35	weakly
2	$V_o(1)$	2	2	o	1	39	weakly
3	$\mathcal{O}(1) + V_o(1)$	3	4	2	o	40	strictly

All of these, except perhaps the case of 39 nodes, are known to Gallarati [6, p.53] . He also gave an example [5] of a sextics with 63 nodes. However, similarly to Togliatti's example above, it is not clear, that more than 24 of its nodes are even.

Remark : After finishing to work on this paper I learnt :

1. there are sextic surfaces with an arbitrarily given number μ of nodes, $1 \leq \mu \leq 64$, see [12] .

2. Beauville's method [3] shows that whenever a sextic surface has μ strictly even nodes, then $\mu = 24, 32, 40$ or 48. It is not known whether $\mu = 48$ actually occurs.

REFERENCES :

[1] BARTH , W. Some properties of stable rank-2 vector bundles on $\mathbb{P}_n$. Math. Ann. 226, 125-150 (1977).

[2] BARTH, W. Kummer surfaces associated with the Mumford Horrocks bundle, talk at Angers (1979).

[3] BEAUVILLE, A. Surfaces with many nodes ($\mu(5) = 31$), talk at Angers (1979).

[4] CATANESE, F. Babbage's conjecture, contact of surfaces, symmetrical determinant varieties and applications, preprint, Pisa, (1979).

[5] GALLARATI, D. Intorno ad una superficie del sesto ordine avente 63 nodi. Boll. U.M.I. Serie III, Anno VII, 392-396 (1952).

[6] GALLARATI, D. Richerche sul contatto di superficie algebriche lungo curve. Mem. Acad. royal Belg. XXXII fasc. 3 (1960).

[7] HARTSHORNE, R. Algebraic Geometry, Springer Verlag (1977).

[8] HARTSHORNE, R. Stable vector bundles of rank 2 on $\mathbb{P}^3$. Math. Ann. 238, 229-280 (1978).

[9] PORTEOUS, I.R. Simple singularities of maps. Springer Lecture Notes in Mathematics 192, 286-307 (1971).

[10] TOGLIATTI, E. Una notevole superficie di 5° ordine con soli punti doppi isolati, Festschrift R. Fueter, Zürich (1940), 127-135.

[11] VOGELAAR, J.A. Constructing vector bundles from codimension-two subvarieties. proefschrift, Leiden (1978).

[12] CATANESE, F. CERESA, G. Constructing sextic surfaces with a given number of nodes, preprint (1979).

Note added in proof :

The generalisation of Porteous' formula to symmetric maps is given in the following preprints :

LORING TU : Variation of Hodge Structure and the Local Torelli Problem. Thesis Harvard 1979.

T. JOZEFIAK, A. LASCOUX, P. PRAGACZ : Classes of determinantal varieties associated with symmetric and antisymmetric matrices.

Wolf BARTH
Mathematisches Institut
Universität Erlangen-Nürnberg
Bismarckstr. 1 $\frac{1}{2}$
D-8520 Erlangen -
West Germany

GROUPE DE JAUGE ÉLARGI
ET
CONNEXIONS STABLES

par

Jean Pierre BOURGUIGNON

Dans cet exposé, nous nous intéressons à certaines formes harmoniques à valeurs dans un fibré, qui apparaissent comme "courbures harmoniques" de connexions.

Bien que nous nous attachions à présenter des résultats récents obtenus à propos de la fonctionnelle de Yang-Mills, nous ignorons délibérément le cadre physique dans lequel cette fonctionnelle s'est révélée intéressante (cf [2]). Cette motivation non proprement mathématique est pourtant sans aucun doute à l'origine de l'engouement (certains diraient de la mode) que l'étude de cette fonctionnelle a suscité ces dernières années.

Dans ce texte nous nous attachons à expliquer en quoi les problèmes posés par cette fonctionnelle sont un prolongement naturel de ceux résolus par la théorie de Hodge. Nous essayons aussi de montrer que les méthodes intrinsèques de la géométrie différentielle et de l'analyse globale se révèlent être des outils adaptés à résoudre certains des problèmes rencontrés.

La théorie ordinaire des formes différentielles harmoniques sert de fil conducteur à tout l'exposé : elle est rappelée au paragraphe 1. Pour définir une différentielle extérieure des formes différentielles extérieures à valeurs dans un fibré, nous introduisons la notion de connexion au paragraphe 2. Dans le paragraphe 3, nous présentons l'approche variationnelle de la théorie des formes harmoniques.

Nous consacrons le paragraphe 4 à la notion, propre à la théorie des formes différentielles extérieures à valeurs dans un fibré, de groupe de jauge et à celle de groupe de jauge élargi. Nous terminons l'exposé en présentant une nouvelle preuve d'un théorème de stabilité des SU_2-connexions sur la sphère standard S^4 (cf[3]), où nous mettons en relief le rôle joué par la non-compacité du groupe des transformations conformes de S^4.

§ 1. FORMES DIFFÉRENTIELLES EXTÉRIEURES HARMONIQUES

Soit (B,g) une variété riemannienne C^∞ de dimension n compacte et α une k-forme différentielle extérieure C^∞ sur B (nous notons $\Omega^k B$ l'espace de ces formes muni de la topologie C^∞). Nous supposerons que α est harmonique, c'est-à-dire que α est fermée (i.e. $d\alpha = 0$ où d désigne la différentielle extérieure) et cofermée (i.e. $\delta^g\alpha = 0$ où δ^g désigne la codifférentielle définie, en utilisant la connexion de Levi-Civita D^g de la métrique g, par

$$\delta^g\alpha(X_1,\ldots,X_{k-1}) = -\sum_{i=1}^{n} (D^g_{e_i}\alpha)(e_i,X_1,\ldots,X_{k-1}),$$

$X_1,\ldots,X_{k-1}$ étant des vecteurs tangents à B et (e_i) une base orthonormée de TB).

La codifférentielle δ^g dépend de la métrique g choisie sur B et n'est rien d'autre que l'adjoint de la différentielle extérieure d pour le produit scalaire sur les k-formes différentielles extérieures $(\alpha,\beta) \mapsto \int_B g(\alpha,\beta)v_g$ où v_g est l'élement de volume défini par la métrique.

Si la variété B est orientée, il est possible de définir l'opérateur de Hodge $*_g$, qui applique $\Lambda^k T^*B$ dans $\Lambda^{n-k}T^*B$, par la formule

$$(*_g\alpha) \wedge \beta = g(\alpha,\beta)\ \omega_g ,$$

où ω_g est l'unique n-forme (induisant l'orientation) qui soit de longueur un.

Il peut sembler au premier abord que $*_g$ ne dépend pas de la métrique g, mais uniquement de ω_g : il n'en est rien, car ω_g détermine seulement une application de $\Lambda^k T^*B$ dans $\Lambda^{n-k}TB$; la métrique est utilisée pour appliquer $\Lambda^{n-k}TB$ dans $\Lambda^{n-k}T^*B$. Lorsque n est pair et $k = \frac{n}{2}$, $*_g$ ne dépend que de la classe conforme de la métrique g (i.e. de la métrique définie à une fonction positive sur B près).

La restriction δ^g_k de la codifférentielle aux k-formes vérifie

$$\delta^g_k = (-1)^{n(k+1)+1} *_g \circ d \circ *_g ,$$

de telle sorte que la k-forme α est harmonique si et seulement si

$$\begin{cases} d\alpha = 0 \\ d*_g\alpha = 0 . \end{cases}$$

§ 2. G-FIBRÉS ET CONNEXIONS

Nous nous intéressons à des fibrés C^∞ de base B, de groupe structural un groupe de Lie compact G d'algèbre de Lie $\mathfrak{g}$. Deux formalismes équivalents permettent d'étudier de tels fibrés : celui des fibrés principaux ou celui des fibrés vectoriels (nous supposons que les représentations de G utilisées sont orthogonales et localement fidèles). Au fibré principal $p : P \to B$ est associé le fibré vectoriel $\pi : E \to B$ où $E = P \times_\rho \xi$ (cf [5] page 55) où ξ est l'espace de la représentation ρ de G.

Un exemple particulièrement intéressant d'une telle représentation est donné par la représentation adjointe ad de G dans son algèbre de Lie $\mathfrak{g}$; le fibré associé est noté $\mathfrak{g}_P \to B$ et a de plus une structure d'algèbre de Lie dans les fibres (on peut aussi considérer le fibré en groupes $G_P = P \times_{Ad} G \to B$ qui est le fibré des automorphismes intérieurs de $p : P \to B$).

Grâce à la différentielle de la représentation ρ, l'algèbre de Lie $\mathfrak{g}$ peut être considérée comme une sous-algèbre de Lie de $\xi^* \otimes \xi$. Ce plongement admet une version fibrée au-dessus de B : nous noterons $\mathfrak{g}_E \to B$ le fibré en sous-algèbres de Lie de $E^* \otimes E \to B$ ainsi obtenu.

Si ω est une k-forme différentielle extérieure sur B à valeurs dans E (nous notons $\Omega^k(B,E)$ l'espace de ces formes en remarquant que $\Omega^k(B, B \times \mathbb{R})$ s'identifie à $\Omega^k B$), nous souhaitons définir une différentielle extérieure "d" de ω. Si nous prenons ω décomposée, i.e. $\omega = \alpha \otimes s$ avec α dans $\Omega^k B$ et s dans $\Omega^o(B,E)$, il est naturel d'imposer que "d" soit une dérivation, i.e. que

$$"d"(\alpha \otimes s) = d\alpha \otimes s + (-1)^k \alpha \wedge "d"s \ .$$

Il est clair que l'opérateur "d" introduit sur les sections de E doit vérifier, pour une fonction f définie sur B,

$$"d"(fs) = f"d"s + df \otimes s \ .$$

De tels opérateurs de $\Omega^o(B,E)$ dans $\Omega^1(B,E)$ sont appelés des connexions.

Si ∇ et ∇' désignent deux connexions, il est facile de voir que

$$(\nabla' - \nabla) fs = f(\nabla' - \nabla) s \ .$$

Ainsi, $\nabla' - \nabla$ étant linéaire par rapport aux fonctions définies sur B , l'espace des connexions sur E apparaît comme un espace affine modelé sur l'espace vectoriel $\Omega^1(B, E^* \otimes E)$.

Il est souvent commode de noter $\nabla_X s$ pour $\nabla s(X)$ (X est un vecteur tangent à B).

Si maintenant nous imposons à la connexion de préserver la G-structure dans les fibres, on vérifie que l'espace des G-connexions sur E est un espace affine modelé sur l'espace vectoriel $\Omega^1(B, \mathfrak{g}_E)$ (par exemple si ∇' et ∇ préservent la métrique dans les fibres, alors $(\nabla' - \nabla)_X$ est un champ d'endomorphismes antisymétriques des fibres).

Une connexion ∇ sur le G-fibré E étant choisie, la formule sur les formes décomposées permet de définir une notion de différentielle extérieure pour les formes différentielles extérieures à valeurs dans E que nous noterons d^∇ .

Il est important de remarquer que, si ∇ est une G-connexion sur E , on peut définir naturellement une G-connexion sur $E^* \otimes E$ (qui est le fibré associé à la représentation $\rho^* \otimes \rho$) en imposant la relation de naturalité

$$(\nabla_X \ell)(s) = \nabla_X(\ell(s)) - \ell(\nabla_X s)$$

où X est un vecteur tangent à B , s une section de E et ℓ une section de $E^* \otimes E$.

La formule donnée au paragraphe 1 pour définir la codifférentielle est encore valable pour les formes à valeurs dans un fibré si on remarque que le fibré $\Lambda^k T^* M \otimes E \to B$ est muni naturellement d'une connexion, par "produit tensoriel" de la connexion induite par la connexion de Levi-Civita de la métrique g sur $\Lambda^k T^* M$ et de la connexion ∇ sur E . Nous notons δ^∇ la codifférentielle ainsi obtenue (nous ne faisons plus figurer la dépendance par rapport à g car nous supposons g fixée dans la suite).

Une k-forme différentielle extérieure ω à valeurs dans E est dite ∇-harmonique si

$$\begin{cases} d^\nabla \omega = 0 \\ \delta^\nabla \omega = 0 \, . \end{cases}$$

Si B est orientée, ce système d'équations aux dérivées partielles est équivalent au système

$$\begin{cases} d^{\nabla}\omega = 0 \\ d^{\nabla} * \omega = 0 \end{cases}$$

(car l'opération $*$ ne fait pas intervenir le fibré E).

§ 3. FORMES HARMONIQUES ET CALCUL DES VARIATIONS

L'importance des formes harmoniques ordinaires vient de ce qu'elles sont les extrêmales d'une fonctionnelle. Sur l'espace affine des k-formes différentielles extérieures α' telles que $\alpha' = \alpha + d\beta$ où α est une k-forme fermée fixée et β une (k-1)-forme quelconque (d'après le théorème de de Rham, nous travaillons donc dans une classe de cohomologie fixée), nous considérons la fonctionnelle

$$F_k(\beta) = \int_B g(\alpha',\alpha')v_g \ .$$

Toutes les formes α' sont fermées car $d^2 = 0$. Il est par ailleurs facile de voir que l'équation $\delta\alpha' = 0$ est l'équation d'Euler de F_k (puisque, comme nous l'avons déjà remarqué, δ est l'adjoint de d). Mieux, F_k n'a qu'une valeur critique, en fait un minimum absolu (cf [9]) ce qui définit β à une différentielle exacte près , car, si on prend α harmonique comme origine affine ,

$$F_k(\beta) = F_k(0) + \int_B g(d\beta,d\beta)v_g \ .$$

Lorsque B est orientée, la fonctionnelle F_k peut aussi être définie par

$$F_k(\beta) = \int_B (\alpha + d\beta)\wedge *(\alpha + d\beta)$$

(en particulier, si B est de dimension paire, $F_{n/2}$ ne dépend que de la classe conforme de g).

Dans ce cas et si α est harmonique, $*\alpha$ est encore fermée en sorte que $F_k(0)$ s'interprète cohologiquement comme

$$F_k(0) = [\alpha]\cup[*\alpha]\ [B]$$

où $\cup$ désigne le cup-produit des classes de cohomologie et [B] la classe fondamentale déduite de l'orientation.

Soulignons que la considération de la fonctionnelle F_k a introduit une dissymétrie entre les deux équations du système définissant les formes harmoniques : l'une est l'équation d'Euler d'un problème du calcul des variations posé sur un espace de formes solutions de l'autre équation.

Remarquons qu'en fait on aurait pu étudier une autre fonctionnelle définie sur un espace de formes cofermées dont les points critiques auraient été des formes fermées, donc harmoniques. Ceci est possible à cause de la linéarité du système d'équations.

Que se passe-t-il dans le cas des formes à valeurs dans un fibré ? D'abord la différentielle extérieure des formes différentielles extérieures à valeurs dans un fibré ne donne pas lieu à un complexe, comme dans le cas du fibré trivial.

Nous posons $R^\nabla = d^\nabla \circ d^\nabla$ que nous appelons courbure de la connexion ∇. Comme $d^2 = 0$, R^∇ n'est pas un opérateur différentiel d'ordre deux, mais d'ordre 0 de telle sorte que R^∇ est en fait un élément de $\Omega^2(B, \mathfrak{g}_E)$ (en effet

$$R^\nabla (fs) = d^\nabla \circ d^\nabla (fs) = d^\nabla (fd^\nabla s) + d^\nabla(df \otimes s)$$
$$= f\, d^\nabla \circ d^\nabla(s) = f\, R^\nabla (s)).$$

De plus la courbure vérifie une identité supplementaire appelée deuxième identité de Bianchi :

$$d^\nabla R^\nabla = 0$$

(en effet, par définition de la connexion induite sur un fibré d'endomorphismes, nous avons

$$d^\nabla R^\nabla = d^\nabla \circ R - R^\nabla \circ d^\nabla = (d^\nabla)^3 - (d^\nabla)^3 = 0).$$

Il n'est donc pas possible de décrire aussi simplement les formes à valeurs dans un fibré appartenant à la même "classe de cohomologie".

La courbure dépend de la connexion par la formule non-linéaire

$$R^{\nabla'} - R^\nabla = d^\nabla A + \frac{1}{2} [A \wedge A]$$

où $\nabla' = \nabla + A$ est une autre connexion (donc A est un élément de $\Omega^1(B, \mathfrak{g}_E)$) et où $[A \wedge A]$ désigne la 2-forme à valeurs dans $\mathfrak{g}_E$ déduite par le crochet de Lie dans les fibres de la 2-forme $A \wedge A$ à valeurs dans $\mathfrak{g}_E \otimes \mathfrak{g}_E$

La seule généralisation des fonctionnelles F_k introduites précédemment à laquelle nous nous intéressons ici est la fonctionnelle de Yang-Mills qui, à une G-connexion ∇ , associe

$$\mathcal{YM}(\nabla) = \int_B g(<R^\nabla, R^\nabla>) v_g$$

où < , > désigne un produit scalaire G-invariant supposé choisi dans les fibres (par exemple l'opposé de la forme de Killing de $\xi^* \otimes \xi$ induite sur $\rho_* \mathfrak{g}$).

Alors que la fonctionnelle F_k était un polynôme inhomogène de degré 2 en la (k-1)-forme β , $\mathcal{YM}(\nabla)$ dépend de la connexion ∇ de façon beaucoup plus compliquée.

Remarquons que, si B est orientée, $\mathcal{YM}$ peut aussi s'exprimer sous la forme

$$\mathcal{YM}(\nabla) = \int_B <R^\nabla \wedge * R^\nabla> .$$

Le principal problème posé par la fonctionnelle de Yang-Mills est de déterminer ses points critiques. Nous nous intéresserons en particulier aux connexions stables, i.e. aux minima locaux de $\mathcal{YM}$.

L'équation des points critiques de $\mathcal{YM}$ est

$$\delta^\nabla R^\nabla = 0 ,$$

car, au premier ordre, la non-linéarité de la fonctionnelle n'apparaît pas. Comme R^∇ est d^∇-fermée par l'identité de Bianchi, nous cherchons donc des 2-formes harmoniques à valeurs dans $\mathfrak{g}_E$ d'un type spécial.

L'analogie avec la fonctionnelle F_2 définie sur une classe de cohomologie de degré 2 est donc assez complète.

Bien entendu nous nous attendons à ce que, dans la théorie de Yang-Mills, non seulement la géométrie de (B,g) mais aussi celle du fibré E interviennent dans la description des points critiques de $\mathcal{YM}$.

Remarquons pourtant que, pour des raisons d'homogénéité, si la dimension de B est supérieure à 4 et si le fibré est trivial en dehors du squelette de codimension 5, alors la borne inférieure de $\mathcal{YM}$ est 0 . Cette borne n'est atteinte, bien sûr, que si la connexion est plate (donc, en passant au besoin à un revêtement de B, si le fibré est trivial). Par contre si B est orientée de dimension 4, il est possible de donner une interprétation cohomologique d'une borne inférieure de $\mathcal{YM}$. En effet $\int_B \langle R^\nabla \wedge R^\nabla \rangle$ est indépendant de la connexion ∇ et s'interprète, à une constante universelle c près, comme le <u>nombre de Pontryaguine</u> $p_1(E)[B]$ du fibré E. Or, si nous décomposons la 2-forme R^∇ (à valeurs dans $\mathfrak{g}_E$) en ses parties positive et négative ($R^\nabla = R^\nabla_+ + R^\nabla_-$), nous obtenons

$$c\,|p_1(E)[B]| = \left|\int_B \langle R^\nabla_+, R^\nabla_+\rangle - \langle R^\nabla_-, R^\nabla_-\rangle\, v_g\right| \leqslant \int_B (\langle R^\nabla_+, R^\nabla_+\rangle + \langle R^\nabla_-, R^\nabla_-\rangle) v_g = \mathcal{YM}(\nabla) .$$

Les minima absolus de $\mathcal{YM}$ s'identifient alors aux connexions ∇ telles que, soit $R^\nabla_- = 0$ (dites <u>connexions autoduales</u>), soit $R^\nabla_+ = 0$ (dites <u>connexions antiautoduales</u>).

§ 4. GROUPE DE JAUGE ET GROUPE DE JAUGE ÉLARGI

Un autre phénomène propre aux formes à valeurs dans un fibré est l'existence d'un groupe non trivial d'automorphismes : <u>le groupe de jauge</u> $\mathcal{G}_E$, groupe des sections du fibré $G_E \to B$. C'est un groupe de dimension infinie qui agit sur l'espace $\mathcal{C}_E$ des connexions sur le fibré E de la façon suivante : pour x dans $\mathcal{G}_E$, ∇ dans $\mathcal{C}_E$ et s dans $\Omega^0(B,E)$,

$$\nabla^x s = (x \circ \nabla \circ x^{-1})(s).$$

On vérifie facilement que

$$R^{\nabla^x} = x \circ R^\nabla \circ x^{-1} .$$

L'espace tangent en la connexion ∇ à l'orbite $\mathcal{G}_E.\nabla$ est décrit, en dérivant la formule précédente, comme image de l'opérateur d^∇ dans $\Omega^1(B,\mathfrak{g}_E)$. L'opérateur d^∇ sur $\Omega^0(B,\mathfrak{g}_E)$ étant à symbole injectif, l'espace $T_\nabla(\mathcal{G}_E.\nabla) = \operatorname{Im} d^\nabla$ est un sous-espace fermé facteur direct dans $\Omega^1(B,\mathfrak{g}_E)$. Un espace transverse à l'orbite est donné par le noyau de l'opérateur δ^∇, adjoint de d^∇ pour le produit scalaire L^2, (pour une situation analogue, voir [4]) puisque

$$T_\nabla \mathcal{C}_E = \Omega^1(B,\mathfrak{g}_E) = \operatorname{Im} d^\nabla \oplus \operatorname{Ker} \delta^\nabla.$$

Il est alors important de noter que la fonctionnelle de Yang-Mills est invariante par le groupe de jauge (en effet dans chaque fibre un élément x de $\mathcal{G}_E$ induit une transformation orthogonale de la métrique fibrée). En conséquence le groupe de jauge agit sur l'ensemble des points critiques de $\mathcal{YM}$(en fait x dans $\mathcal{G}_E$ applique les formes ∇-harmoniques dans les formes ∇^x -harmoniques).

Dans la théorie de Hodge, le groupe I(B,g) des isométries de la métrique g opère dans l'espace de toutes les formes harmoniques, la composante connexe de l'identité opérant trivialement. De plus, lorsque B est orientée de dimension paire, les $\frac{n}{2}$-formes harmoniques ne dépendent que de la classe conforme de la métrique et sont donc invariantes par le groupe C(M,g) des transformations conformes de la métrique.

Nous appellerons groupe de jauge élargi $\tilde{\mathcal{G}}_E$ le groupe des automorphismes de E qui revêtent une isométrie de (B,g) lorsque $n \neq 4$ ou une transformation conforme de (B,g) si n = 4 et B est orientée. Nous avons une suite exacte

$$0 \to \mathcal{G}_E \to \tilde{\mathcal{G}}_E \to \mathcal{J}(B,g) \to 0$$

où $\mathcal{J}(B,g) = \mathbf{I}(B,g)$ si $n \neq 4$ ou $\mathcal{J}(B,g) = C(B,g)$ si $n = 4$.

Il est intéressant de décrire l'espace tangent à l'orbite $\tilde{\mathcal{G}}_E.\nabla$ au point ∇ . Pour cela nous utilisons la construction géométrique suivante : si X est un champ de vecteurs sur B ,nous construisons $\tilde{X}$ le relèvement ∇-horizontal de X pour la G-connexion ∇ . Le flot de $\tilde{X}$ sur E est formé d'automorphismes du fibré qui induisent sur B le flot de X .Lorsque X est une isométrie ou une transformation conforme (suivant que la dimension de B est ou non égale à 4), $\tilde{X}$ appartient à l'algèbre de Lie du groupe de jauge élargi. Il faut prendre garde que l'application que nous venons de construire n'est pas un homomorphisme d'algèbres de Lie bien qu'elle soit une scission de la suite exacte des espaces tangents en l'identité déduite de la suite exacte précédente. On vérifie alors facilement que

$$T_\nabla (\tilde{\mathcal{G}}_E.\nabla) = \operatorname{Im} d^\nabla + \{ i_X R^\nabla \mid X \in \mathfrak{J}(B,g)\} ,$$

où $\mathfrak{J}(B,g)$ désigne l'algèbre de Lie du groupe $\mathcal{J}(B,g)$.

Soulignons qu'il ne s'agit pas nécessairement d'une somme directe. En effet si nous évaluons $\delta^\nabla(i_X R^\nabla)$ avec X dans $\mathfrak{J}(B,g)$, nous trouvons

$$\delta^{\nabla}(i_X R^{\nabla}) = g(d\xi, R^{\nabla})$$

où ξ désigne la 1-forme duale pour la métrique g du champ de vecteurs X . Ainsi, si ξ est fermée, l'orbite du groupe de jauge élargi est certainement plus grande que celle du groupe de jauge (si $n \neq 4$, ξ est duale d'une isométrie infinitésimale et ξ est fermée si et seulement si elle est parallèle ; si $n = 4$, il existe sur la sphère standard des transformations infinitésimales conformes X telles que ξ soit fermée, voir §5 pour les détails).

§ 5. UN THÉORÈME DE STABILITÉ

Nous nous proposons de prouver le

THÉORÈME (cf [3]).- Toute connexion stable sur un SU_2-fibré sur la sphère S^4 munie de la métrique standard g_o est autoduale ou antiautoduale, donc un minimum absolu de la fonctionnelle de Yang-Mills.

Nous esquissons ici une variante de la preuve donnée dans [3] qui a l'avantage d'être conceptuellement plus claire que celle-ci ; elle fait référence à la notion de groupe de jauge élargi.

Nous utilisons de façon essentielle(*) la partie non compacte du groupe conforme de la sphère standard (d'après [6] ou [7] , S^n est la seule variété compacte dont le groupe conforme est non compact) . Rappelons qu'en dimension 4, le groupe de jauge élargi est une extension du groupe de jauge par le groupe des transformations conformes de la base.

Nous partons de la formule de la dérivée seconde de la fonctionnelle $\mathcal{YM}$ en un point critique :

$$\frac{d^2}{dt^2}\mathcal{YM}(\nabla^t)\Big|_{t=0} = \int_B g(\langle \mathcal{B}^{\nabla} A, A\rangle) v_g$$

où $A = \frac{d}{dt}\nabla^t\Big|_{t=0}$ et

$$\mathcal{B}^{\nabla} A = \delta^{\nabla} d^{\nabla} A + \sum_{i=1}^{n} [R^{\nabla}_{e_i,\cdot}, A_{e_i}] .$$

Le groupe de jauge laissant la fonctionnelle $\mathcal{YM}$ invariante, $\mathcal{YM}$ est constante le long des orbites. Nous pouvons donc travailler transversalement à celles-ci : nous supposerons donc que $\delta^{\nabla} A = 0$ et nous prendrons comme opérateur de la variation

(*) A notre grande surprise, cela n'est en fait pas important comme nous l'expliquons dans le § 10 de [3 bis] écrit en septembre 1979. En conséquence nous pouvons étendre le théorème de stabilité à tous les espaces homogènes orientables, le champ pouvant se réduire à un champ abélien si l'espace a de la 2-cohomologie.

seconde

$$\mathcal{S}^\nabla A = (d^\nabla \delta^\nabla + \delta^\nabla d^\nabla)A + \sum_{i=1}^{n} [R^\nabla_{e_i,\cdot}, A_{e_i}] \quad ,$$

qui a l'avantage d'être elliptique.

La stabilité de la connexion ∇ se traduit par la condition suivante sur $\mathcal{S}^\nabla$: pour tout A dans $\Omega^1(B, \mathfrak{g}_E)$ tel que $\delta^\nabla A = 0$,

$$\int_B g(<\mathcal{S}^\nabla A, A>) v_g \geqslant 0 \quad .$$

Remarquons alors que, si X est une transformation infinitésimale conforme, son relèvement ∇-horizontal $\tilde{X}$ dans l'espace total du fibré appartient à l'algèbre de Lie du groupe de jauge élargi. On vérifie facilement que la variation infinitésimale de connexion induite par $\tilde{X}$ n'est autre que $i_X R^\nabla$ en tant que 1-forme à valeurs dans $\mathfrak{g}_E$. Par suite nous avons nécessairement $\tilde{\mathcal{S}}^\nabla(i_X R^\nabla) = 0$ (en fait même $\mathcal{S}^\nabla(i_X R^\nabla) = 0$, puisqu'on vérifie que $\delta^\nabla(i_X R^\nabla) = 0$). Cette relation se vérifie de façon directe analytiquement mais, à cause de son caractère tautologique (elle traduit l'invariance conforme de la fonctionnelle $\mathcal{YM}$), elle ne donne aucune information sur le champ R^∇. Pour pouvoir obtenir de l'information supplémentaire sur R^∇ nous allons utiliser qu'en dimension 4 justement toute 2-forme harmonique telle que le champ de Yang-Mills R^∇ donne naissance à deux 2-formes harmoniques : ses parties positive et négative R^∇_+ et R^∇_-. Nous allons appliquer l'hypothèse de stabilité à la famille de variations infinitésimales de la connexion $i_X R^\nabla_\pm$ où X varie parmi les transformations infinitésimales conformes.

Nous aurons besoin d'une description précise de l'algèbre de Lie $\mathfrak{C}(S^n, g_o)$ du groupe conforme de la sphère standard. Nous avons

$$\mathfrak{C}(S^n, g_o) = \mathfrak{so}_{n+1} \oplus \mathfrak{PC}$$

où $\mathfrak{so}_{n+1}$ est l'algèbre de Lie du groupe des isométries de (S^n, g_o) et où l'espace vectoriel $\mathfrak{PC}$ des champs de vecteurs purement conformes s'identifie à l'espace vectoriel de dimension n+1 des gradients des premières harmoniques sphériques (ce sont les restrictions à S^n, plongée dans R^{n+1} de façon standard, des fonctions linéaires sur R^{n+1}).

Remarquons que cet espace vectoriel $\mathfrak{PC}$ est muni d'une métrique naturelle puisqu'il s'identifie au dual de R^{n+1} : nous notons $S\mathfrak{PC}$ la sphère de rayon 1 de cet espace vectoriel euclidien.

Revenons au calcul proprement dit. Nous devons évaluer le laplacien de Hodge $d^{\nabla}\delta^{\nabla} \neq \delta^{\nabla}d^{\nabla}$ de la 1-forme $i_X\varphi$ où φ est une 2-forme harmonique à valeurs dans $\mathfrak{g}_E$ (nous prendrons ultérieurement $\varphi = R^{\nabla}_{-}$). Il est commode pour cela d'utiliser la formule de Weitzenböck reliant le laplacien de Hodge au laplacien brut $\nabla^*\nabla$, ce dernier opérateur ayant l'avantage d'être défini sur tous les champs de tenseurs. Il est clair que

$$\nabla^*\nabla\,(i_X\varphi) = i_{D^*DX}\varphi - \sum_{i=1}^{4} i_{D_{e_i}X}(\nabla_{e_i}\varphi) + i_X(\nabla^*\nabla\varphi)$$

où D est la connexion de Levi-Civita de la base et (e_i) une base orthonormée.

En utilisant les propriétés du champ de vecteurs X et la formule de Weitzenböck reliant $\nabla^*\nabla$ et $d^{\nabla}\delta^{\nabla} + \delta^{\nabla}d^{\nabla}$ sur les 2-formes (à valeurs vectorielles) et en écrivant que φ est harmonique, nous obtenons en chaque point m de la sphère

$$\mathcal{S}^{\nabla}(i_X\varphi)_Y = -\sum_{j=1}^{4}\{[R^{\nabla}_{e_j,Y}, \varphi_{e_j,X}] + [R^{\nabla}_{e_j,X}, \varphi_{e_j,Y}]\}.$$

Il est alors clair que la moyenne de la forme quadratique $X \mapsto \langle \mathcal{S}^{\nabla}(i_X\varphi), i_X\varphi\rangle$ sur la sphère de l'espace $\mathbb{P}\mathbb{C}$ est nulle. En effet nous pouvons prendre pour base orthonormée de $\mathbb{P}\mathbb{C}$ une base (ε_j) $j = 0,\dots,4$ telle que $\varepsilon_o(m) = 0$ et $\varepsilon_j(m)=e_j$ pour $j = 1,\dots,4$ de telle sorte que

$$\int_{S\mathbb{P}\mathbb{C}} \langle \mathcal{S}^{\nabla}(i_X\varphi), i_X\varphi\rangle\, d\mu_X = \sum_{i,j,k=1}^{4} \langle [R^{\nabla}_{e_j,e_i}, \varphi_{e_j,e_k}] + [R^{\nabla}_{e_j,e_k}, \varphi_{e_j,e_i}], \varphi_{e_i,e_k}\rangle$$

$$= 0,$$

puisque φ_{e_j,e_k} est antisymétrique en j et k alors que la somme des crochets est symétrique en j et k. Si nous intégrons l'égalité précédente sur la sphère S^4, d'après l'interprétation de $\mathcal{S}^{\nabla}$ comme opérateur de la variation seconde et l'hypothèse de stabilité, nous obtenons pour chaque X dans $\mathbb{P}\mathbb{C}$ $\mathcal{S}^{\nabla}(i_X\varphi) = 0$ dès que φ est une 2-forme harmonique à valeurs dans $\mathfrak{g}_E$. Si nous appliquons celà à $\varphi = R^{\nabla}_{-}$, nous avons, pour tous vecteurs tangents X et Y à S^4,

$$\sum_{j=1}^{4}\{[R^{\nabla}_{+e_j,X}, R^{\nabla}_{-e_j,Y}] + [R^{\nabla}_{+e_j,Y}, R^{\nabla}_{-e_j,X}]\} = 0.$$

Il faut alors utiliser un lemme algèbrique qui établit que le produit tensoriel de l'espace des 2-formes positives par l'espace des 2-formes négatives est isomorphe en tant que SO_4-module avec l'espace des 2-tenseurs symétriques à trace nulle. Les deux termes de droite dans la formule ci-dessus se doublent donc et l'équation précédente implique en fait que, pour tous vecteurs tangents X, Y, Z et T a S^4, nous avons

$$[R^{\nabla}_{+Y,Z}, R^{\nabla}_{-T,X}] = 0 .$$

Le théorème suit finalement après avoir étudié le commutateur d'un élément de la forme $R^{\nabla}_{+X,Y}$ dans la fibre de $\mathcal{G}_E$ (c'est pourquoi le théorème est encore vrai pour les SU_3-fibrés, mais ne l'est pas pour les autres groupes simples sans hypothèse supplémentaire), car nous obtenons que soit R^{∇}_+, soit R^{∇}_- est nul.

Si nous pouvions prouver que, sans l'hypothèse de stabilité sur ∇ ,

$$\delta^{\nabla}(i_X R^{\nabla}_+) = 0 \qquad \text{pour tout } X \text{ de } \mathfrak{FK} ,$$

alors nous aurions montré que les seuls points critiques de la fonctionnelle $\mathcal{YM}$ sur l'espace des SU_2-connexions des fibrés sur S^4 sont les minima absolus, i.e. les connexions autoduales ou antiautoduales (pour une description complète de ces connexions, voir [1]). Le problème de l'existence d'autres points critiques semble, à nos yeux, un des problèmes majeurs soulevés par l'étude de la fonctionnelle $\mathcal{YM}$ sur la sphère S^4.

RÉFÉRENCES

[1] M. F. ATIYAH, N. HITCHIN, N. J. DRINFELD, Y. I. MANIN, Physics Lett. A, 65 (1978), 185-187.

[2] J. P. BOURGUIGNON, Geometry and physics of Yang-Mills fields, Proc. Congress Math. Society of Japan, Nagoya (1979).

[3] J. P. BOURGUIGNON, H. B. LAWSON, J. SIMONS, Stability and gap phenomena for Yang-Mills fields, Proc. Nat. Acad. Sci. USA, 76 (1979), 1550-1553.

[3 bis] J.P. BOURGUIGNON, H.B. LAWSON, Stability and isolation phenomena for Yang-Mills fields, soumis à Comm. Math. Physics.

[4] D. G. EBIN, The manifold of Riemannian metrics, Proc. Symp. Pure Math. A.M.S. XV, Berkeley, (1970), 11-40.

[5] S. KOBAYASHI, K. NOMIZU, Foundations of differential geometry, Interscience (1963).

[6] J. LELONG-FERRAND, Transformations conformes et quasi-conformes des variétés riemanniennes compactes, Mém. Acad. Roy. Belg. Cl. Sci., 39 (1971).

[7] M. OBATA, The conjectures on conformal transformations of Riemannian manifolds, J. of Diff. Geom. 6 (1971), 247-258.

[8] G. de RHAM, Variétés différentiables, Hermann (1955).

[9] F. WARNER, Foundations of differentiable manifolds and Lie groups, Scott, Foresman and Co (1971).

J. P. BOURGUIGNON
Centre de Mathématiques
Ecole Polytechnique
91128 Palaiseau Cedex (France)

(Laboratoire associé au C.N.R.S. n° 169)

LES FIBRES UNIFORMES DE RANG AU PLUS n SUR $\mathbb{P}_n(\mathbb{C})$ SONT CEUX QU'ON CROIT.

G. ELENCWAJG*
A. HIRSCHOWITZ,
M. SCHNEIDER*

INTRODUCTION.

Les fibrés dont il s'agit sont des fibrés vectoriels algébriques.

Le résultat que nous démontrons est si crédible que plusieurs experts (cf. [11] [12] [13]) ont cru le lire dans un travail (Sato [7]) où il ne figure pas :

Théorème :

Tout fibré uniforme de rang r sur $\mathbb{P}_n(\mathbb{C})$, avec $r \leq n$, est d'un des deux types suivants :

- somme directe de fibrés en droites
- produit tensoriel par un fibré en droites du fibré tangent ou du fibré cotangent (dans ce cas, r égale n).

Rappelons qu'un fibré uniforme est un fibré dont les restrictions aux droites projectives sont isomorphes deux à deux. Pour r égale à 2, le résultat est connu depuis VAN DE VEN [9], tandis que SATO [7] en a démontré deux autres cas particuliers importants ($r < n$ et n impair). Notre progrès résulte peut-être d'un examen systématique de l'obstruction topologique à l'existence de fibrés uniformes de type de scindage donné (équation $\mathscr{E}$ du § 4).

* *Ce travail a été effectué pendant que les auteurs extrêmes étaient enseignants associés à l'Université de Nice.*

On peut espérer un résultat analogue si $r=n+1$ (pour $n=2$, $r=3$, cf. ELENCWAJG [2]). En revanche, on sait (cf. ELENCWAJG [3]) qu'il existe des fibrés uniformes non homogènes de rang 4 sur $\mathbb{P}_2(\mathbb{C})$ et pour n au moins égal à 3, nous donnons en Appendice une construction très simple de fibrés uniformes et non homogènes de rang 3n-1 sur $\mathbb{P}_n(\mathbb{C})$.

Nous remercions bien cordialement Joel BRIANÇON et Jean Michel GRANGER qui ont bien voulu démontrer pour nous le lemme (7.2).

§ 1. LA CONSTRUCTION STANDARD.

Dans ce paragraphe on associe à tout fibré uniforme divers entiers et des fibrés sur une variété grassmannienne.

1.1. Fibrés sur une droite projective : On sait (au moins depuis GROTHENDIECK [5]) qu'à tout fibré A sur une droite projective L, on peut associer des entiers $(k\,;\mu_1,\ldots,\mu_k\,;\,r_1,\ldots,r_k)$ vérifiant $\mu_1>\ldots>\mu_k$ et $r_1>0,\ldots,r_k>0$ de sorte que A soit isomorphe à $\bigoplus_{i=1}^{k} r_i\,\mathcal{O}_L(\mu_i)$. Il en découle que A est muni d'une filtration naturelle $HN^{\cdot}(A)$ (c'est la filtration de HARDER-NARASIMHAN [10]) : HN^jA est l'unique sous fibré de A isomorphe à $\bigoplus_{i=1}^{j} r_i\,\mathcal{O}_L(\mu_i)$. On peut aussi le décrire par la formule :

$$HN^jA = \operatorname{Im}\left[H^0(L,A(-\mu_j))\otimes\mathcal{O}_L(\mu_j)\longrightarrow A\right].$$

1.2. Fibrés sur une droite projective relative : la construction précédente s'étend au cas avec paramètre sous une hypothèse d'uniformité :

Proposition (1.2.) :

Soit G une variété algebrique munie d'un fibré D de rang deux et soit $q : F \longrightarrow G$ le Proj associé[1)] à D. Soit A un fibré sur F tel que les nombres

1) La fibre de F est l'espace projectif des droites de la fibre de D.

$k(g)$, $\mu_i(g)$ et $r_i(g)$ associés aux restrictions $A(g)$ de A aux fibres $q^{-1}(g)$ de q soient indépendants du point g de G. Alors il existe des fibrés $A_1,\dots,A_k$ sur G de rang $r_1,\dots,r_k$ et une filtration $HN_q^{\bullet}A$ par des sous-fibrés de A de gradué associé

$$\bigoplus_{i=1}^{k} [q^*(A_i) \otimes \mathcal{O}_F(\mu_i)] .$$

Démonstration :

Précisons d'abord que, par $\mathcal{O}_F(\ell)$ on désigne comme d'habitude la puissance tensorielle ℓ-ième du dual du fibré de Hopf (relatif) associé à D.

On prend pour k ; $\mu_1,\dots,\mu_k$; $r_1,\dots,r_k$ les valeurs communes $k(g)$; $\mu_1(g),\dots,\mu_k(g)$; $r_1(g),\dots,r_k(g)$. La filtration est la filtration de Harder- Narasimhan relative à q avec

$$HN_q^j A = \operatorname{Im} [q^*q_*(A \otimes \mathcal{O}_F(-\mu_j))\cdot \otimes \mathcal{O}_F(\mu_j) \longrightarrow A].$$

Si on pose

$$A_j = q_*[(gr^j\, HN_q^{\bullet}A) \otimes \mathcal{O}_F(-\mu_j)]$$

(où gr^j désigne le j-ième terme de l'objet gradué associé), d'après le théorème de changement de base, $HN_q^j A$ et A_j sont des fibrés qui ont les propriétés requises. C.Q.F.D.

<u>1.3. Notations</u> : Tout au long de cet article, on étudie un fibré uniforme E de rang r sur l'espace projectif $\mathbb{P}_n(\mathbb{C})$ des droites de $\mathbb{C}^{n+1}$. On désigne par G la grassmannienne des droites de $\mathbb{P}_n(\mathbb{C})$ (donc des plans de $\mathbb{C}^{n+1}$) et par F la variété d'incidence :

$$F: = \{(x,L) \in \mathbb{P}_n \times G \mid x \in L \ \}.$$

On note p et q les projections de F sur $\mathbb{P}_n$ et G. On pose $D := q_* p^* \mathcal{O}_{\mathbb{P}_n}(1)$. C'est un fibré de rang deux dont F est le Proj, avec $\mathcal{O}_F(1) = p^* \mathcal{O}_{\mathbb{P}_n}(1)$. Comme E est uniforme, la proposition (1,2) s'applique au fibré $A := p^* E$ pour donner le résultat qu'on retiendra du § 1 ;

Corollaire (1.3.) :

A tout fibré uniforme E sur $\mathbb{P}_n$ sont (naturellement) associés des entiers $(k;\mu_1 > \ldots > \mu_k ; r_1,\ldots,r_k)$ et des fibrés $E_1,\ldots,E_k$ de rangs $r_1,\ldots,r_k$ sur G tels que p^*E admette une filtration (par des sous-fibrés) de gradué associé

$$\bigoplus_{i=1}^{k} [q^*(E_i) \otimes p^* \mathcal{O}_{\mathbb{P}_n}(\mu_i)] .$$

§ 2. LA VARIETE D'INCIDENCE F.

2.1. Quelques fibrés tautologiques : notons

M le sous-fibré tautologique de rang 2 sur G,

N le fibré quotient tautologique de rang n-1 sur G,

H le sous-fibré tautologique (de Hopf) de rang 1 sur $\mathbb{P}_n$,

et Q le fibré quotient tautologique de rang n sur $\mathbb{P}_n$.

On utilisera l'isomorphisme naturel entre le fibré tangent et Hom(H,Q) : grâce à lui, on peut, dans l'énoncé du théorème, remplacer les fibrés tangent et cotangent par Q et Q^*.

Comme Proj D, Proj M et Proj Q s'identifient naturellement à F. Parmi les quatre fibrés tautologiques correspondants, seul le fibré de Hopf relatif à Q mérite une nouvelle notation, soit H_Q. En écrivant les suites exactes d'Euler relatives à M et Q et en appliquant la formule donnant l'espace tangent relatif d'un Proj, on obtient la

Proposition (2.1.) :

a) Il existe deux suites exactes (naturelles) :

$$0 \longrightarrow p^*H \longrightarrow q^*M \longrightarrow H_Q \longrightarrow 0$$

$$0 \longrightarrow H_Q \longrightarrow p^*Q \longrightarrow q^*N \longrightarrow 0$$

b) les espaces tangents relatifs sont donnés par

$$T_{F/G} = H_Q \otimes p^*H^*, \; T_{F/\mathbb{P}_n} = q^*N \otimes H_Q^*.$$

2.2. Cohomologie de F : Introduisons les classes de Chern $U := c_1(p^*H)$ et $V := c_1(H_Q)$, ainsi que le polynôme $R(X,Y) := X^n + \ldots + X^\ell Y^{n-\ell} + \ldots + Y^n$.

La proposition suivante décrit l'anneau de cohomologie de F en relation avec ceux de $\mathbb{P}_n$ et G :

Proposition (2.2) :

Soit π le morphisme naturel de $\mathbb{Z}[U,V]$ dans $H^*(F,\mathbb{Z})$.

a) π est surjectif ; son noyau est l'idéal engendré par $R(U,V)$ et U^{n+1} (autrement dit, π identifie $H^*(F,\mathbb{Z})$ à $\mathbb{Z}[U,V] / (R(U,V),U^{n+1})$).

b) La sous-algèbre $p^*H^*(\mathbb{P}_n,\mathbb{Z})$ est l'image par π de l'algèbre des polynômes indépendants de V.

c) La sous-algèbre $q^*H^*(G,\mathbb{Z})$ est l'image par π de l'algèbre des polynômes symétriques en U et V.

Démonstration :

a) et b) résultent du théorème de Leray-Hirsch appliqué au fibré Q grâce à l'isomorphisme naturel entre $(\mathbb{Z}[U]/(U^{n+1}))[V]\ /(R(U,V))$ et $\mathbb{Z}[U,V]\ /\ (U^{n+1},R(U,V))$.

Pour prouver c), on utilise la suite exacte (cf.2.1.) :

$$0 \longrightarrow p^{*}H \longrightarrow q^{*}M \longrightarrow H_Q \longrightarrow 0$$

qui implique

$$q^{*}(c_1(M)) = U+V \text{ et } q^{*}(c_2(M)) = UV.$$

Comme la cohomologie de G est engendrée par $c_1(M)$ et $c_2(M)$ (cf. e.g. DOLD [4]) , $q^{*}\ H^{*}(G,\mathbb{Z})$ est engendré par les polynômes en U+V et UV, c'est à dire (cf. NEWTON [6]) par les polynômes symétriques en U et V. C.Q.F.D.

2.3. Groupe de Picard de F.

Proposition (2.3) :

Le groupe de Picard de F est le groupe libre (commutatif) engendré par $p^{*}H$ et H_Q.

Démonstration :

Soit L un fibré inversible sur F. D'après (2.2), il existe un fibré L' de la forme $p^{*}H^{a} \otimes H_Q^{b}$ vérifiant $c_1(L) = c_1(L')$. D'après le théorème de changement de base, $L'L^{-1}$ est isomorphe à $p^{*}p_{*}(L'L^{-1})$. Il en résulte que $c_1(p_{*}L'L^{-1})$ est nul, donc que $p_{*}(L'L^{-1})$ est trivial et $L'L^{-1}$ aussi.
C.Q.F.D.

§ 3. REDUCTION A LA TOPOLOGIE ALGEBRIQUE.

3.1. L'énoncé de topologie algébrique : l'objet de ce paragraphe est de ramener la preuve du théorème à celle de l'énoncé suivant qui sera démontré dans les paragraphes ultérieurs :

Proposition T :

Soit E un fibré topologique de rang r sur $\mathbb{P}_n$. Soient k ; $\mu_1,\ldots,\mu_k$; $r_1,\ldots,r_k$ des entiers vérifiant $r_1 \geq 1,\ldots,r_k \geq 1$. Soient $E_1,\ldots,E_k$ des fibrés topologiques de rangs $r_1,\ldots,r_k$ sur G. On suppose que p^*E admet une filtration de gradué associé

$$\bigoplus_{i=1}^{k} \left[(q^*E_i) \otimes p^* \mathcal{O}_{\mathbb{P}_n}(\mu_i)\right].$$

On fait l'hypothèse supplémentaire : (H) n majore r et si n égale r, les μ_i sont consécutifs décroissants.

Alors seuls les cas suivants sont possibles :

(C1) $k=1$

(C2) $c_1(E_1) = \ldots = c_1(E_k) = 0$

(C3) $r_1=1$ et $q^*c_1(E_1) = U+V$

(C3)* $r_k=1$ et $q^*c_1(E_k) = -U-V$.

Pour mener à bien la réduction du théorème à la proposition T, il nous faut justifier l'hypothèse (H) et démontrer le théorème moyennant chacune des hypothèses C1, C2, C3, C3*.

3.2. L'hypothèse (H) :

les extensions de fibrés en droites sur $\mathbb{P}_n$ étant triviales, la proposition suivante justifie l'hypothèse H :

Proposition (3.2.) : (cf. VAN DE VEN [9], ELENCWAJG [2], SPINDLER [8],...) : supposons, avec les notations de 1.3 que pour un indice j avec $1 \leq j < k$, on ait $\mu_{j+1} \leq \mu_j - 2$. Alors E est extension de deux fibrés uniformes de rangs $r_{j+1}+\ldots+r_k$ et $r_1+\ldots+r_j$.

Démonstration :

Le sous-fibré $HN_q^j\, p^*E$ définit un $\mathbb{P}_n$-morphisme ε de $p : F \longrightarrow \mathbb{P}_n$ dans la grassmannienne relative $g : Gr_s\, E \longrightarrow \mathbb{P}_n$ des sous-espaces vectoriels de dimension $s := r_1+\ldots+r_j$ de E. Si W désigne le sous-fibré tautologique (de rang s) de g^*E, on sait que l'espace tangent relatif de cette grassmannienne s'identifie à $\mathcal{H}om(W, g^*E/W)$. Par ailleurs ε est caractérisé par l'égalité $\varepsilon^*W = HN_q^j\, p^*E$. La différentielle relative de ε est donc une section du fibré $\mathcal{H}om(T_{F/\mathbb{P}_n}, \varepsilon^*\mathcal{H}om(W, g^*E/W))$ c'est à dire du fibré $\mathcal{H}om(T_{F/\mathbb{P}_n}, \mathcal{H}om(HN_q^j\, p^*E, p^*E/HN_q^j p^*E))$. D'après (2.1.b) ce fibré s'écrit encore

$$q^*N^* \otimes H_Q \otimes \mathcal{H}om(HN_q^j p^*E,\ p^*E/HN_q^j\, p^*E).$$

D'après (2.1.a) la restriction de H_Q à une fibre L de q a pour première classe de Chern 1. La restriction à L du fibré ci-dessus est donc isomorphe à $(n-1)\mathcal{O}_L(1) \otimes \mathcal{H}om(\bigoplus_{i \leq j} r_i\, \mathcal{O}_L(\mu_i), \bigoplus_{i > j} r_i\, \mathcal{O}_L(\mu_i))$ et a zéro pour seule section. Sa différentielle relative étant nulle, ε est constant sur les fibres de p et l'image dans $\mathbb{P}_n \times Gr_s E$ de son graphe est le graphe d'une section algébrique de $Gr_s E$. Soit E' le sous-fibré correspondant de E, et $E'' := E/E'$ le quotient correspondant. Par construction, on a $p^*E' = HN_p^j\, p^*E$ et $p^*E'' = p^*E/HN_q^j p^*E$ ce qui montre que E' et E'' sont uniformes. C.Q.F.D.

3.3. Le cas C1 :

Il est connu depuis VAN DE VEN [9].

Nous le déduirons de la

Proposition (3.3) :

Soient E et K des fibrés respectivement sur $\mathbb{P}_n$ et G. Si p^*E et q^*K sont isomorphes alors E et K sont triviaux.

Démonstration :

Si E est trivial, K, qui est isomorphe à q_*p^*E, l'est aussi.

Soit D le carré cartésien de $\mathbb{P}_n$ avec les deux projections pr_1 et pr_2 et soit $\pi : \hat{D} \longrightarrow D$ l'éclatement de la diagonale dans D. Au dessus de D $(= \mathbb{P}_n \times \mathbb{P}_n)$, on a deux plongements naturels de $\hat{D}$ dans $F \times \mathbb{P}_n$ et $\mathbb{P}_n \times F$ et les morphismes correspondants $\hat{pr}_i : \hat{D} \longrightarrow F$ (pour i=1,2) vérifient

$po\hat{pr}_i = pr_i o\pi$ et $qo\hat{pr}_1 = qo\hat{pr}_2$:

$$\begin{array}{ccccc} G & \xleftarrow{q} & F & \overset{\hat{pr}_1}{\underset{\hat{pr}_2}{\leftleftarrows}} & \hat{D} \\ & & \Big\downarrow p & & \Big\downarrow \pi \\ & & \mathbb{P}_n & \overset{pr_1}{\underset{pr_2}{\leftleftarrows}} & D \end{array}$$

Pour que E soit trivial, il (faut et il) suffit que pr_1^*E et pr_2^*E soient isomorphes ce qui, puisque $\pi_* \mathcal{O}_{\hat{D}}$ égale $\mathcal{O}_D$, est le cas dès que $\pi^* pr_1^* E$ et $\pi^* pr_2^* E$ le sont. Or $\pi^* pr_i^* E$ est isomorphe à $\hat{pr}_i^* q^* K$ et la conclusion résulte de l'égalité entre $qo\hat{pr}_1$ et $qo\hat{pr}_2$. C.Q.F.D.

Corollaire (3.3) :

Soit E un fibré uniforme sur $\mathbb{P}_n$. Si k(E) égale 1, alors E est isomorphe à $r_1\mathcal{O}(\mu_1)$.

Démonstration :

Il suffit d'appliquer la proposition précédente à $E(-\mu_1)$ et E_1. C.Q.F.D.

3.4. Le cas C2 :

Dans le cas où n égale 2 (et k égale 2), si $c_1(E_1)$ et $c_1(E_2)$ sont nuls, E_1 et E_2 sont triviaux pour des raisons évidentes. Pour le cas général (n quelconque) on utilise le

Lemme (3.4.) :

Soit K un sous-fibré d'un fibré trivial sur $\mathbb{P}_{n-1}$. Si $c_1(K)$ est nul, alors K est trivial.

Démonstration :

La restriction de K à une droite L est isomorphe à $\underset{i}{\oplus}\, \mathcal{O}_L(a_i)$ avec les a_i négatifs ou nuls et de somme nulle. Ces a_i sont donc tous nuls et le corollaire précédent s'applique. C.Q.F.D.

Proposition (3.4.) :

Soit E un fibré uniforme sur $\mathbb{P}_n$. Si les premières classes de Chern des fibrés associés (cf. 1.3), $c_1(E_1),\dots,c_1(E_k)$ sont nulles, alors E est somme directe de fibrés en droites.

Démonstration :

Si k égale 1, la proposition (3.3) s'applique. On peut donc raisonner par récurrence.

Le sous-fibré $q^*E_1 \otimes p^*\mathcal{O}_{\mathbb{P}_n}(\mu_1)$ de p^*E a, d'après le lemme, une restriction triviale sur chaque fibre de p.

D'après le théorème de changement de base, il existe donc un sous-fibré E' de E vérifiant $HN^1_q p^* E = p^* E'$. Le quotient $E'' := E/E'$ est donc uniforme avec pour $1 \leq i \leq k-1$, $E''_i = E_{i+1}$. L'hypothèse de récurrence s'applique à E'' et E' (pour E' on peut aussi appliquer 3.3). Il reste pour conclure à observer que les extensions de fibrés en droites sur $\mathbb{P}_n (n \geq 2)$ sont triviales. C.Q.F.D.

3.5. Les cas C3 et C3* :

Ils se traitent ensemble car ils sont "duaux" l'un de l'autre. L'argument central est constitué par la

Proposition (3.5.) : (cf. VAN DE VEN [9]) :

Soit $p : F \longrightarrow X$ le Proj d'un fibré Q et H_Q le fibré de Hopf relatif. Pour qu'un fibré E sur X admette un sous-fibré isomorphe à Q, il (faut et il) suffit que $p^* E$ admette un sous-fibré isomorphe à H_Q.

Démonstration :

On sait que le morphisme naturel de $p^* Q^*$ dans H_Q^* induit un isomorphisme entre Q^* et $p_* H_Q^*$. En tensorisant par $p^* E$, on obtient que le morphisme naturel de $\mathcal{H}om(p^* Q, p^* E)$ dans $\mathcal{H}om(H_Q, p^* E)$ induit un isomorphisme entre $\mathcal{H}om(Q,E)$ et $p_* \mathcal{H}om(H_Q, p^* E)$. Par ailleurs il est évident que, dans cet isomorphisme, les injections se correspondent. C.Q.F.D.

Corollaire (3.5) :

Soit E un fibré uniforme sur $\mathbb{P}_n$.

- si $r_1 = 1$ et $q^* c_1(E_1) = U+V$, alors E admet un sous-fibré isomorphe à $Q(\mu_1 - 1)$.
- si $r_k = 1$ et $q^* c_1(E_k) = -U-V$, alors E admet un quotient isomorphe à $Q^*(\mu_k + 1)$.

Démonstration :

Le second cas se ramène au premier en considérant E^*. Dans le premier cas, on observe que $q^*E_1 \otimes p^* \mathcal{O}_{\mathbb{P}_n}(\mu_1)$ a pour première classe de Chern $V+(1-\mu_1)U$ qui est aussi la première classe de Chern du fibré de Hopf relatif associé à $Q(\mu_1-1)$ et (2.3) permet d'appliquer la proposition précédente.
C.Q.F.D.

Remarque :

Sato [7] énonce un résultat plus général qui permettrait d'éviter l'hypothèse H et le § 3.2.

§ 4. L'EQUATION $\mathcal{E}$ ET LE CAS a=0.

4.1. Polynôme de Chern :

A tout fibré topologique E de rang r sur X, associons le polynôme (de Chern)[1] :

$$C_E(T) := T^r - c_1(E)\, T^{r-1} + \ldots + (-1)^r\, c_r(E).$$

Les règles de calcul des classes de Chern se traduisent de la façon suivante :

a) Si L est un fibré en droites sur X, alors

$$C_{E\otimes L}(T) = C_E(T - c_1(L)).$$

b) Si E est filtré de sorte que le gradué associé soit $\bigoplus_i E_i$, alors

$$C_E = \prod_i C_{E_i}.$$

Dans les hypothèses de la proposition (3.1), C_{p^*E} admet un représentant privilégié dans $\mathbb{Z}[T,U,V]$, celui qui est indépendant de V (et homogène) ; nous le notons E(T,U). De même, si k diffère de 1, le fibré E_i est de rang au plus n-1 et $C_{q^*E_i}$ admet dans $\mathbb{Z}[T,U,V]$ un unique représentant homogène de degré r_i que nous notons $E_i(T,U,V)$.

1) ce n'est pas exactement la définition usuelle.

4.2. Le nombre a :

On se place encore sous les hypothèses de la proposition T avec k différent de 1. D'après 4.1., les polynômes $E(T,U)$ et $\pi_i E_i(T+\mu_i U,U,V)$ ont même image dans $H^*(F,\mathbb{Z})[T]$. Comme ils sont de degré au plus n, il existe une unique constante a: = a(E) telle que

$$E(T,U) - \pi_i E_i(T+\mu_i U,U,V) = aR(U,V).$$

4.3. Le cas a=0 :

Il se produit en particulier si r est strictement inférieur à n et donne lieu à la

Proposition (4.3.) :

Soit E comme dans la proposition T. Si a(E) est nul, alors

$$c_1(E_1) = \ldots = c_1(E_k) = 0$$

Démonstration :

Le polynôme E_i divise $E(T-\mu_i U,U,V)$ donc est indépendant de V et (par symétrie) de U. C.Q.F.D.

Corollaire (4.3) (Sato [7]) :

Tout fibré uniforme de rang strictement inférieur à n sur $\mathbb{P}_n$ est une somme directe de fibrés en droites.

Remarque (pour Hartshorne) :

La démonstration du cas particulier précédent[1)] n'utilise que les § 1, § 2,§3.3 et § 3.4. En particulierelle n'utilise pas le § 3.2, ce qui est satisfaisant puisque Sato démontre cet énoncé sur un corps quelconque.

4.4. Vocabulaire :

On appellera solution de $\mathcal{E}$ tout quintuplet (P ; a ; k ; S ; µ) où

1) C'est essentiellement celle de Sato.

- P est un polynôme à deux variables homogènes de degré n, à coefficients entiers :

$$P(T,U) = \sum_{\ell=0}^{n} p_\ell \, U^\ell \, T^{n-\ell}$$

monique en T (i.e. $p_o=1$) ;

- a est un entier ;
- k est un entier au moins égal à 1 ;
- $S=(S_1,\ldots,S_k)$ est une suite de k polynômes à trois variables (T,U,V) à coefficients entiers, de degrés $(r_1,\ldots,r_k)$ au moins égaux à 1, symétriques en U et V et moniques en T ;
- $\mu = (\mu_1,\ldots,\mu_k)$ est une suite strictement décroissante de k entiers ;

assujettis à l'égalité

$$(\mathcal{E}) \quad \boxed{P(T,U) - aR(U,V) = \prod_{i=1}^{k} S_i(T+\mu_i U,U,V)}$$

Si de plus a est non nul et les μ_i consécutifs, nous dirons que (P ; a ; k ; S ; μ) est <u>solution de $\mathcal{E}^*$</u>.

<u>4.5. Réduction à l'algèbre</u> :

Soit E comme dans la proposition T. Le cas où a est nul est traité dans l'alinéa 4.3. Si a est non nul, alors r égale n et (c'est l'hypothèse H), les μ_i sont consécutifs. Pour prouver la proposition T, il nous suffit par conséquent de démontrer l'énoncé algebrique suivant :

Proposition A :

Soit (P ; a ; k ; S ; μ) une solution de $\mathcal{E}^*$:

1°) Si $p_n=0$, alors $k=1$.

2°) Si $p_n\neq 0$, alors $k=2$ et $S_1=T-U-V$ ou $S_2=T+U+V$.

4.6. Vocabulaire :

La démonstration de la proposition précédente occupe les trois paragraphes suivants et passe par l'étude des "solutions approchées".

Nous appelons solution approchée (de $\mathcal{E}^*$) tout polynôme $P = P(T,U) = \Sigma\, p_\ell\, U^\ell\, T^{n-\ell}$ à deux variables homogène de degré n, à coefficients rationnels tel que P(T,U)-R(U,V) admette pour diviseur un polynôme homogène en (T,U,V) à coefficients rationnels, symétrique en U et V, de degré au moins égal à 1. Si (P ; a ; k ; S ; μ) est une solution de $\mathcal{E}^*$, pour $i=1,\ldots,k$, $P_i(T,U) := \frac{1}{a} P(T-\mu_i U,U)$ est une solution approchée dont on dira qu'elle est associée à la solution donnée.

§ 5. LES SOLUTIONS APPROCHEES.

Dans ce paragraphe, on étudie les solutions approchées P à travers l'intersection dans $\mathbb{P}_2(\mathbb{C})$ de la courbe d'équation P(T,U) = R(U,V) avec la droite d'équation T=0. On met en évidence deux espèces de solutions approchées et on montre qu'il nous suffit, pour chaque espèce, de déterminer les solutions approchées dites primitives.

5.1. Notations :

On note D_o la droite de $\mathbb{P}_2(\mathbb{C})$ d'équation T=0 sur laquelle on distingue la coordonnée inhomogène V/U ce qui permet d'identifier $\mathbb{C}$ à un ouvert de D_o. On note P une solution approchée et C la courbe d'équation P(T,U) = R(U,V).

5.2. L'intersection $C \cap D_o$:

Elle est essentiellement constituée de racines de l'unité :

Proposition (5.2.) :

Deux cas seulement sont possibles

1°) $p_n = 0$ et alors $C \cap D_o$ est constitué par les n racines (n+1)-ièmes de l'unité différentes de 1.

2°) $p_n = 1$ et alors $C \cap D_o$ contient, outre 0, les (n-1) racines n-ièmes de l'unité différentes de 1.

Démonstration :

Soit S_o un diviseur symétrique non constant de P(T,U)-R(U,V).
En faisant T=0, on constate que $S_o(0,U,V)$ divise $p_n\ U^n-R(U,V)$. Par symétrie $S_o(0,U,V)$ divise aussi $p_n\ V^n-R(U,V)$. Par soustraction, $S_o(0,U,V)$ divise donc $p_n(U^n-V^n)$.

1°) Si $p_n = 0$, une équation de $C \cap D_o$ est $R(U,V) = 0$.

2°) Si p_n est non nul, $S_o(0,U,V)$ divise les deux polynômes $F: = p_n\ U^n-R(U,V)$ et $G: = U^n-V^n$, donc aussi $(U-V)F+VG = (p_n-1)U^n(U-V)$. Comme les seuls diviseurs symétriques de $U^n(U-V)$ sont les constantes, p_n égale 1. Par suite une équation de $C \cap D_o$ est
$V(U^{n-1} +\ldots+ U^{k-1}\ V^{n-k} +\ldots+ V^{n-1}) = 0.$ C.Q.F.D.

5.3. Vocabulaire :

Nous dirons que P est une solution approchée de première espèce si p_n est nul et de seconde espèce sinon. Nous dirons que P est une solution approchée primitive si on peut choisir le diviseur symétrique S_o de sorte que $S_o(0,1,z)$ admette pour zéro une racine $(n-p_n+1)$-ième de l'unité primitive.

5.4. Réduction au cas primitif.

Proposition (5.4.) :

L'une au moins des solutions approchées associées à une solution de $\mathcal{E}^{*}$ est primitive.

Démonstration :

Soit (P ; a ; k ; S ; µ) une solution de $\mathcal{E}^{*}$.

Si k égale 1, la solution approchée associée est évidemment primitive.

Si n et k sont égaux à 2, alors $S_i(T,U,V)$ est de la forme $T-b_i(U+V)$ avec b_i non nul et les deux solutions approchées associées sont évidemment primitives.

Si n égale 3, alors k est au plus égal à 2 car U^2+V^2 et U^2+UV+V^2 sont irréductibles sur $\mathbb{Z}$; si k égale 2, l'un des r_i égale 2 et la solution approchée associée correspondante est primitive.

On va donc supposer n au moins égal à 4 et utiliser les inégalités :

$$0 < \frac{\pi}{n} < \frac{2\pi}{n+1} < \frac{2\pi}{n} < \frac{3\pi}{n} < \frac{4\pi}{n+1} < \pi \quad .$$

Pour la première solution approchée, qui vérifie

$$P_1(T,U) - R(U,V) = \frac{1}{a} \prod_i S_i(T+(\mu_i-\mu_1)U,U,V),$$

la proposition (5.2) assure l'existence d'un indice ℓ tel que $S_\ell(\mu_\ell-\mu_1,\ 1,\ \exp\frac{2i\pi}{n})$ ou $S_\ell(\mu_\ell-\mu_1,\ 1,\ \exp\frac{2i\pi}{n+1})$ soit nul. Nous allons prouver que S_ℓ $(0,1,\ \exp\frac{2i\pi}{n})$ ou $S_\ell(0,1,\ \exp\frac{2i\pi}{n+1})$ est nul.

Pour cela, il nous suffit de démontrer que $S_\ell(0,1,z)$ a une racine non nulle d'argument compris entre $\frac{\pi}{n}$ et $\frac{3\pi}{n}$. Introduisons à cet effet l'ensemble K des réels t tels que $S_\ell(t-\mu_1,1,z)$ ait une racine non nulle r(t) admettant un argument strictement compris entre $\frac{\pi}{n}$ et $\frac{3\pi}{n}$. Comme K contient μ_ℓ et est manifestement ouvert, il suffit de montrer que K est fermé. Soit donc t_o adhérant à K et z_o une racine de $S_\ell(t_o-\mu_1,1,z)$ adhérant à tous les ensembles $r(K \cap]t_o-\varepsilon,\ t_o+\varepsilon[)$. Comme, pour t dans K,

$S_\ell(t-\mu_1,1,z)$ est divisible par $(z-r(t))(z-\overline{r(t)})$ et que $P(t_o,1)-R(1,z)$ n'est pas divisible par z^2, la racine z_o est non nulle. Il nous suffit dès lors de prouver que pour m égal à 1 ou 3 et d réel, $R(1,z)+d$ n'a pas de racine non nulle d'argument $\frac{m\pi}{n}$, ou encore que, pour ρ strictement positif, $R(1,\rho \exp \frac{im\pi}{n}) + d$ ne s'annule pas. Après multiplication par $\rho \exp \frac{im\pi}{n} - 1$, il ne reste qu'à observer que

$$\rho^{n+1} \exp \frac{im(n+1)\pi}{n} - 1 + d\rho \exp \frac{im\pi}{n} - d \text{ est égal à}$$

$$\rho(d-\rho^n) \exp \frac{im\pi}{n} - (1+d)$$

qui ne s'annule pas. C.Q.F.D.

§ 6. LES SOLUTIONS APPROCHEES PRIMITIVES.

L'objet de ce paragraphe est de montrer que si $(P;a;k;S;\mu)$ est solution de $\mathcal{E}^*$, P est de la forme $R(T-\nu U,\rho U)$. Le résultat découle de la classification des solutions approchées primitives qui s'obtient en étudiant (au voisinage infinitésimal d'une racine primitive de l'unité dans D_o) l'intersection de la courbe C avec la droite D_x d'équation $T=x(U+V)$.

6.1. Première espèce :

Proposition (6.1.) :

Les seules solutions approchées primitives de première espèce sont les monômes bT^n.

Démonstration :

Soit P une solution approchée primitive de première espèce, S_o un diviseur symétrique de degré maximum de $P(T,U) - R(U,V)$ et y_o une racine $(n+1)$-ième primitive de l'unité vérifiant $S_o(0,1,y_o) = 0$. On note y la solution holomorphe de $S_o(x(1+y(x)),1,y(x)) = 0$ vérifiant $y(0) = y_o$.

L'existence et l'unicité de ce germe de fonction sont assurées par le théorème des fonctions implicites car y_o est racine simple de R(1,z) donc de $S_o(0,1,z)$.

Nous allons montrer par récurrence que pour m=1,...,n-1, on a

$$y^{(m)}(0) = p_{n-m}=0.$$

En observant que S_o divise P(T,U)-R(U,V) et son symétrique, on obtient les relations suivantes :

(1) $$P(x(1+y),1) - R(y,1) = 0$$

(2) $$P(x(1+y),y) - R(y,1) = 0.$$

Calculons les dérivées en 0 :

(1') $$p_{n-1}(1+y_o) - R'y'(0) = 0$$

(2') $$p_{n-1}(1+y_o)\ y_o^{n-1} - R'y'(0) = 0$$

où R' vérifie $(n+1)y_o^n = R'(y_o-1) + R(y_o,1) = R'(y_o-1)$.

On constate donc que $R'(1+y_o)(y_o^{n-1}-1)$ est non nul d'où on déduit que p_{n-1} et y'(0) sont nuls ce qui amorce la récurrence.

Grâce à l'hypothèse de récurrence, les dérivées m-ièmes en 0 des équations (1) et (2) s'écrivent :

(1^m) $$m!\ p_{n-m}(1+y_o)^m - R'\ y^{(m)}(0) = 0$$

(2^m) $$m!\ p_{n-m}(1+y_o)^m\, y_o^{n-m} - R'\ y^{(m)}(0) = 0$$

et le fait que y_o soit racine (n+1)-ième primitive de l'unité permet de conclure. C.Q.F.D.

<u>6.2. Seconde espèce</u> :

Proposition (6.2.) :

Les seules solutions approchées primitives de seconde espèce sont les polynômes R(bT,U).

Démonstration :

Soit $\Sigma_+(T,U,V)$ un diviseur symétrique en U et V de $P(T,U) - R(U,V)$ et y_o une racine n-ième primitive de l'unité vérifiant $\Sigma_+(0,1,y_o) = 0$.

On observe que $R(p_{n-1}T,U)$ est aussi solution approchée et que la relation

$R(p_{n-1}T,U) - R(U,V) = (p_{n-1}\ T-V)\ \ \Sigma_-(T,U,V)$

définit un polynôme Σ_- symétrique en U et V et vérifiant

$\Sigma_-(0,1,y_o) = 0$. On va montrer que $P_+ := P(T,U)$ et $P_- := R(p_{n-1}T,U)$ sont égaux. A cet effet, comme dans la démonstration précédente, on introduit la solution holomorphe $y_\pm$ du système :

$$\Sigma_\pm(x(1+y_\pm(x)),\ 1,\ y_\pm(x)) = 0$$
$$y_\pm(0) = y_o.$$

et on a les relations :

$$(1_\pm)\qquad P_\pm(x(1+y_\pm),1) - R(y_\pm,1) = 0$$
$$(2_\pm)\qquad P_\pm(x(1+y_\pm),y_\pm) - R(y_\pm,1) = 0.$$

On va montrer par récurrence sur m que p_{n-m} égale p_{n-1}^m et que $y_+^{(m)}(0)$ et $y_-^{(m)}(0)$ sont égaux.

En dérivant $(1_\pm)$ en 0, on obtient

$$p_{n-1}(1+y_o) - R'y_\pm'(0) = 0$$

avec, cette fois, $(y_o-1)R' + \dfrac{y_o^{n+1}-1}{y_o-1} = (n+1)y_o^n$, soit $R' = \dfrac{n}{y_o-1}$.

Et, R' étant non nul, cette relation amorce la récurrence.

Pour $2 \leqslant m \leqslant n$, posons $p_m^+ = p_m$ et $p_m^- = p_{n-1}^{n-m}$.

Si, dans les dérivées m-ièmes en 0 des équations $(1_\pm)$ et $(2_\pm)$, on néglige les termes qui, par hypothèse de récurrence, sont indépendants du signe $\pm$, on obtient :

$(1^m_\pm)$ $$m!\ p^{\pm}_{n-m}(1+y_o)^m - R'\ y^{(m)}_{\pm}(0) + \ldots = 0$$

$(2^m_\pm)$ $$m!\ p^{\pm}_{n-m}(1+y_o)^m\ y_o^{n-m} + ny_o^{n-1}\ y^{(m)}_{\pm}(0) - R'y^{(m)}_{\pm}(0) + \ldots = 0$$

En multipliant la seconde équation par y_o et en observant que $n-R'y_o$ égale $-R'$, on constate que $(p^{\pm}_{n-m}, y^{(m)}_{\pm}(0))$ est solution d'un système (2-2) indépendant du signe ±, de déterminant $R'm!(1+y_o)^m\ (y_o^{n-m+1} - 1)$. Cela permet de conclure puisque $n-m+1$ est compris entre 1 et $n-1$ et que y_o est racine primitive de l'unité. C.Q.F.D.

§ 7. LES SOLUTIONS DE $\mathcal{E}^*$.

Pour achever la démonstration de la proposition A et donc celle du théorème, il nous reste à traiter les deux cas issus des propositions (6.1) et (6.2).

7.1. Première espèce :

Proposition (7.1.) :

Si $((T-\nu U)^n\ ;\ a\ ;\ k\ ;\ S\ ;\mu\ )$ est solution de $\mathcal{E}^*$, alors k égale 1.

Démonstration :

Les coefficients non dominants du polynôme en T : $T^n - aR(U,V)$ sont divisible par le polynôme irréductible, $V - U \exp \frac{2i\pi}{n+1}$. D'après le critère d'Eisenstein, le terme constant n'étant pas divisible par le carré de $V - U \exp \frac{2i\pi}{n+1}$, $T^n - aR(U,V)$ est irréductible. Par changement de variable, il en est de même pour $(T-\nu U)^n - aR(U,V)$. C.Q.F.D.

7.2. Seconde espèce :

Il nous faut mieux connaître le polynôme $\Sigma_n(T,U,V) := \sum_{t+u+v=n} T^t U^u V^v$,

qui vérifie $(T-V)\,\Sigma_n(T,U,V) = R(T,U) - R(U,V)$.

La démonstration du lemme qui suit a été trouvée pour nous par Joel Briançon et Jean-Michel Granger :

Lemme (7.2.) :

$\Sigma_n(T,U,V)$ est irréductible.

Démonstration :

On montre par récurrence sur n que le cône Γ_n d'équation $\Sigma_n = 0$ dans $\mathbb{C}^3 - \{0\}$ est lisse (ce qui permet de conclure d'après le théorème de Bezout). Le cas n=1 étant évident, nous allons montrer que si Γ_n est singulier en (t,u,v), Γ_{n-1} aussi.

On remarque les identités :

(1) $\left(\frac{\partial}{\partial T} + \frac{\partial}{\partial U} + \frac{\partial}{\partial V}\right) \Sigma_n = (n+2)\,\Sigma_{n-1}$

(2) $\Sigma_n = T\,\Sigma_{n-1} + R(U,V) = U\,\Sigma_{n-1} + R(V,T) = V\,\Sigma_{n-1} + R(T,U)$

De (1) on déduit $\Sigma_{n-1}(t,u,v) = 0$ et de (2), $t^{n+1} = u^{n+1} = v^{n+1}$ ($\neq 0$).

En dérivant (2), on obtient :

$$\frac{\partial}{\partial T}\Sigma_n = \Sigma_{n-1} + T\frac{\partial}{\partial T}\Sigma_{n-1}$$

$$\frac{\partial}{\partial U}\Sigma_n = \Sigma_{n-1} + U\frac{\partial}{\partial U}\Sigma_{n-1}$$

$$\frac{\partial}{\partial V}\Sigma_n = \Sigma_{n-1} + V\frac{\partial}{\partial V}\Sigma_{n-1}$$

D'où $\frac{\partial}{\partial T}\Sigma_{n-1}(t,u,v) = \frac{\partial}{\partial U}\Sigma_{n-1}(t,u,v) = \frac{\partial}{\partial V}\Sigma_{n-1}(t,u,v) = 0$. C.Q.F.D.

Proposition (7.3.) :

Si $(R(T,pU) ; a ; k ; S ;\mu)$ est une solution de $\mathcal{E}^*$ avec p entier non nul, alors k égale 2 et p^2 égale 1.

En outre si p égale 1, alors on a $\mu_1 = 1$ et $S_1 = T-U-V$,

si p égale - 1, alors on a $\mu_2 = -1$ et $S_2 = T+U+V$.

Démonstration :

On sait que a égale p^n. On a

$R(T,pU) - p^n R(U,V) = R(T,pU) - R(pV,pU) = (T-pV)\ \Sigma_{n-1}(T,pU,pV)$.

De l'irréductibilité de Σ_{n-1} (et par factorialité), on déduit $k \leq 2$.

Et k diffère de 1 parce que $R(T-\nu U,pU)$ n'est symétrique pour aucune valeur de ν . Comme Σ_{n-1} est symétrique et que T-pV égale T+pU - p(U+V), on voit que les valeurs prises par μ sont 0 et p.

Comme elles sont consécutives, deux cas seulement sont possibles :

$p=1,\ \mu_1=1,\ S_1=T-U-V$,

$p=-1, \mu_2=-1\ S_2=T+U+V$. C.Q.F.D.

APPENDICE

On sait que sur $\mathbb{P}_2$, il existe des fibrés uniformes de rang 4 et plus qui ne sont pas homogènes (cf. Elencwajg [3]) et donc qu'on ne peut obtenir par des constructions algebriques (somme, produit tensoriel, symétrique, extérieur) à partir des fibrés tautologiques.

Ce phénomène se reproduit sur $\mathbb{P}_n$ pour $n \geq 3$: nous construisons ici (pour $n \geq 3$) un fibré de rang $3n-1$ sur $\mathbb{P}_n$, uniforme et non homogène.

Nous reprenons les notations des § 1 et § 2 et nous notons E_L la restriction du fibré E à la droite L.

A1 Construction de fibrés uniformes :

Soit K un fibré uniforme avec $k(K) \geq 2$ et $\mu_k(K) = 0$. Alors toute section s de K induit, via la projection de p^*K sur q^*K_k, une section s de K_k. Pour construire et étudier de nouveaux fibrés uniformes, on dispose de la

Proposition (A1) :

Soit K uniforme avec $k(K) \geq 2$ et $\mu_k(K) = 0$. Soit $s : \mathcal{O} \longrightarrow K$ une section de K telle que s ne s'annule pas. Si $\pi : K \longrightarrow E$ désigne le conoyau de s, alors E est uniforme. En outre, pour toute droite L, π induit un isomorphisme entre HN^1K_L et HN^1E_L.

Démonstration :

Soit φ un isomorphisme entre K_L et $\underset{i}{\oplus}\, r_i\, \mathcal{O}_L(\mu_i)$. Si on décompose $\varphi \circ s_{|L}$ sous la forme $(t_1,\ldots,t_k)$, l'hypothèse assure que t_k ne s'annule pas. Si S est un supplémentaire (trivial) de $t_k\, \mathcal{O}_L$ dans $r_k\, \mathcal{O}_L(\mu_k)$ $(= r_k\, \mathcal{O}_L)$, alors $\varphi^{-1}\left[\underset{i<k}{\oplus}\, r_i\, \mathcal{O}_L(\mu_i) \oplus S\right]$ est un supplémentaire de $s\mathcal{O}_L$ dans K_L.

Par suite, E_L est isomorphe à $\underset{i<k}{\oplus}\, r_i\, \mathcal{O}_L(\mu_i) \oplus (r_k-1)\mathcal{O}_L$, E est uniforme et pour $j < k$, π induit un isomorphisme entre $HN^j K_L$ et $HN^j E_L$. C.Q.F.D.

A.2. Défaut d'homogéneité :

Pour montrer que le fibré E que nous construisons n'est pas homogène, nous montrerons qu'il n'est pas "biuniforme" (cf. Daoudi [1]) : pour tout fibré E sur $\mathbb{P}_n$ et tout couple (L',L") de droites distinctes sécantes, on note $\delta_E(L',L")$ la dimension de l'intersection des fibres en $L' \cap L"$ de $HN^1\, E_{L'}$ et $HN^1\, E_{L"}$. Comme le groupe projectif opère transitivement sur les couples de droites distinctes sécantes, on a le résultat suivant (démontré sous une forme plus générale dans [1]).

Proposition (A.2.) :

Si E est homogène, δ_E est constant.

A.3. Compléments sur Q :

Soit x un vecteur non nul de $\mathbb{C}^{n+1}$. On lui associe un point $\bar{x}$ de $\mathbb{P}_n$ et des sections s_x et t_x des quotients tautologiques Q et N sur $\mathbb{P}_n$ et G. Avec la notation de A.1., $\bar{s}_x$ et t_x sont égales d'où il résulte :

A.3a : $\bar{s}_x$ ne s'annule que sur les droites passant par $\bar{x}$.

Si L est une droite de $\mathbb{P}_n$, $HN^1 Q_L$ s'identifie au fibré quotient tautologique de rang 1 sur L puisque $HN^1_q Q$ est égal à H_Q(cf. 2.1). Il en résulte les deux remarques suivantes que nous utiliserons pour le calcul de $\delta_E(L',L'')$: soient L' et L" deux droites distinctes se coupant en z. Alors :

__A.3b__. Les fibres de $HN^1Q_{L'}$ et $HN^1Q_{L''}$ en z sont d'intersection nulle.

__A.3c__. $s_x(z)$ est engendré par ces deux fibres si et seulement si $\bar{x}$ est dans le plan engendré par L' et L".

__A.4. Le fibré E :__

On considère trois vecteurs x_1,x_2,x_3 non coplanaires de $\mathbb{C}^{n+1}$ et s_1,s_2,s_3 les sections associées de Q. On considère aussi trois copies Q^1, Q^2, Q^3 de Q et on pose $K: = Q^1 \oplus Q^2 \oplus Q^3$. C'est un fibré uniforme avec $k=2$, $\mu_k=0$. De plus K_k est la somme directes de trois copies de N. Si on pose $s: = (s_1,s_2,s_3)$, alors $\bar{s}$ est égal à $(\bar{s}_1,\bar{s}_2,\bar{s}_3)$ et ne s'annule pas d'après A.3.a. D'après A.1, le fibré $E: = K/s\,\mathcal{O}$ est uniforme et, pour toute droite L, HN^1E_L est isomorphe (via $\pi : K \longrightarrow E$) à HN^1K_L, c'est-à-dire à $\bigoplus_i HN^1\, Q^i_L$.

Soient alors L' et L" deux droites distinctes se coupant en z. D'après A.3b l'intersection des fibres en z de $\bigoplus_i HN^1\, Q^i_{L'}$ et $\bigoplus_i HN^1\, Q^i_{L''}$ est réduite à zéro. Par suite $\delta_E(L',L'')$ est nul sauf si la fibre en z de Ker π est contenue dans $\bigoplus_i (HN^1\, Q^i_{L'} \oplus HN^1\, Q^i_{L''})$ auquel cas $\delta_E(L',L'')$ vaut 1. Ceci se produit, d'après A.3c, si et seulement si $\bar{x}_1,\bar{x}_2,\bar{x}_3$ engendrent le même plan que L' et L", ce qui prouve que δ_E prend deux valeurs, 1 et 0.

__Ajouté en mars 1980__ : Philippe ELLIA (Nice) a construit, en perfectionnant la méthode ci-dessus, des fibrés uniformes non homogènes de rang 2n+1 sur $\mathbb{P}_n$.

B I B L I O G R A P H I E

[1] DAOUDI M. Fibrés biuniformes. Preprint Nice.

[2] ELENCWAJG G. Les fibrés uniformes de rang 3 sur $\mathbb{P}_2(\mathbb{C})$ sont homogènes. Math. Ann. 231, 217-227 (1978).

[3] ELENCWAJG G. Des fibrés uniformes non homogènes Math. Ann. 239 185-192 (1979).

[4] DOLD A. Lectures on algebraic topology. Springer Verlag, 1972.

[5] GROTHENDIECK A. Sur la classification des fibrés holomorphes sur la sphère de Riemann. Amer. J. Math. 79 (1957) 121-138.

[6] NEWTON I. The mathematical papers of Isaac Newton.Cambridge the University Press (1967-1976).

[7] SATO E. Uniform vectorbundle on a projective space. J. Math. Soc. Japan 28 123-132 (1976).

[8] SPINDLER H. Der Satz von Grauert-Mülich für beliebige... Math. Ann. 243 131-141 (1979).

[9] VAN DE VEN A. On uniform vector bundles. Math. Ann. 195 245-248 (1972).

[10] HARDER G. NARASIMHAN M.S. On the cohomology groups of moduli spaces of vector bundles on curves. Math. Ann. 212 215-248 (1975)

[11] BARTH W. In Math. Reviews 53, 2952

[12] HARTSHORNE R. Algebraic Vector Bundles on Projective Spaces : a problem list. Topology 18 117-128 (1979).

[13] SCHNEIDER M. Holomorphic vector bundles on $\mathbb{P}_n$. Sem. Bourbaki n° 530 (Nov. 1978).

TYPE DE SCINDAGE GÉNÉRALISÉ POUR LES FIBRÉS STABLES.

O. FORSTER,
A. HIRSCHOWITZ,
M. SCHNEIDER.

INTRODUCTION.

Dans un récent travail, SPINDLER [15] a démontré le résultat suivant : Si E est un fibré semi-stable sur $\mathbb{P}_n(\mathbb{C})$, alors sa restriction à une droite générale est de la forme $E|L \cong \mathcal{O}_L(a_1) \oplus \ldots \oplus \mathcal{O}_L(a_r)$, avec $a_i \geq a_{i+1} \geq a_i - 1$. (La suite $a_1 \geq a_2 \geq \ldots \geq a_r$ s'appelle type de scindage générique de E). La méthode de démonstration raffine une idée de VAN DE VEN [17], qui traitait le cas des fibrés uniformes de rang 2, idée déjà reprise par GRAUERT-MÜLICH [6], BARTH [2], ELENCWAJG [3]. Elle consiste à montrer que si E est un fibré de type de scindage générique $a_1 \geq a_2 \geq \ldots \geq a_r$ avec $a_i - a_{i+1} \geq 2$ pour au moins un i, il existe un sous-faisceau $F \subset E$ de type de scindage générique $a_1 \geq \ldots \geq a_i$.
Le point de départ du présent travail est ce théorème de SPINDLER que nous étendons au cas d'une variété complexe X quelconque. Les droites sont remplacées par des courbes et la filtration de HARDER-NARASIMHAN [8] généralise le type de scindage. On obtient comme corollaire que la famille des faisceaux semi-stables sur X avec polynôme de HILBERT fixé est limitée. Ce résultat était connu dans des cas particuliers (TAKEMOTO [16], MARUYAMA [10], [14], GIESEKER [5], ELENCWAJG-FORSTER [4]) et a des conséquences concernant les espaces de modules
(MARUYAMA [12], [13], LANGTON [9]).

§ 1. CONVENTIONS ET RAPPELS.

Par variété on entend toujours une variété algebrique projective ou quasi-projective sur $\mathbb{C}$. Les notions de sous-ensembles ouverts ou fermés sont prises au sens de Zariski. Un faisceau est toujours un faisceau algébrique cohérent. Un fibré vectoriel est un faisceau localement libre. Un sous-fibré d'un fibré vectoriel est un sous-faisceau qui est localement facteur direct.

Soit maintenant X une variété lisse et E un faisceau sans torsion sur X. Il existe un fermé $A \subset X$ de codimension ≥ 2 en dehors duquel E est un fibré. On appelle rang de E le rang r de ce fibré. On note det E le fibré en droites sur X (déterminé à un isomorphisme près) qui est isomorphe à $\Lambda^r E$ au dessus de $X \setminus A$.

Pour tout sous-faisceau F de E il existe un ouvert dense $U \subset X$ au dessus duquel F est facteur direct. Soit $F' := E/F$ et $F'' := F'/T(F')$, où $T(F')$ est le sous-faisceau de torsion de F'. Alors

$$\hat{F} := \mathrm{Ker}\,(E \longrightarrow F'')$$

est facteur direct de E en dehors d'un fermé de codimension 2 et $\hat{F}$ coïncide sur U avec son sous-faisceau F.

Supposons maintenant X projective, munie d'un fibré en droites très ample $\mathcal{O}_X(1)$. Le degré (relatif à $\mathcal{O}_X(1)$) d'un faisceau E sans torsion est défini comme suit : Soit $D \in |\det E|$ et $H \in |\mathcal{O}_X(1)|$ (si L est un fibré en droites on note $|L|$ l'ensemble de diviseurs qui lui est associé).

Alors $\deg(E) := D.\,H^{n-1}$, $n := \dim X$,

où D. H^{n-1} désigne le nombre d'intersection. Si $H_1,\dots,H_{n-1} \in |\mathcal{O}_X(1)|$ se coupent transversalement suivant une courbe lisse $C \subset X$, alors

$$\deg(E) = \deg(E|C).$$

Si $E \neq 0$, on définit

$$\mu(E) = \frac{\deg(E)}{\mathrm{rang}(E)}$$

E est dit semi-stable (au sens de MUMFORD-TAKEMOTO, cf. [16]), si pour tout sous-faisceau non nul F de E on a

$$\mu(F) \leq \mu(E).$$

C'est équivalent au fait que pour tout quotient sans torsion G de E on a

$$\mu(E) \leq \mu(G).$$

GIESEKER [5] et MARUYAMA [11] utilisent une notion plus restrictive de semi-stabilité.

Soient $F_1 \subset F_2$ deux sous-faisceaux de E qui coïncident au dessus d'un ouvert dense. Alors $\deg(F_1) \leq \deg(F_2)$.

Nous utiliserons aussi le fait suivant : Soit E un faisceau sur une variété X quelconque et F un sous-faisceau de la restriction de E à un ouvert $U \subset X$. Alors il existe un sous-faisceau (cohérent !) de E prolongeant F.

§ 2. FIBRÉS SUR LES COURBES.

Dans tout ce paragraphe, X désigne une courbe projective lisse et E un fibré vectoriel de rang $r > o$ sur X.

D'après HARDER-NARASIMHAN [8] il existe un sous-faisceau unique F de E vérifiant

i) $\mu(F) \geq \mu(F')$ pour tout sous-faisceau $F' \subset E$.

ii) Si $\mu(F) = \mu(F')$ pour un sous-faisceau $F' \subset E$, alors F' est contenu dans F.

Ce faisceau F est un sous-fibré semi-stable de E. Nous l'appelerons sous-fibré semi-stable maximal de E.

Le fibré E possède une filtration canonique

$$0 = E_o \subset E_1 \subset \ldots \subset E_k = E$$

caractérisée par le fait que E_i / E_{i-1} est le sous-fibré semi-stable maximal de E / E_{i-1} pour $i = 1,\ldots,k$. Nous l'appelerons filtration de HARDER-NARASIMHAN.

Si on pose

$$\mu_i : = \mu_i(E) : = \mu(E_i / E_{i-1}),$$

on a

$$\mu_1 > \mu_2 > \ldots > \mu_k.$$

On pose aussi

$$\mu_{max}(E) : = \mu_1, \ \mu_{min}(E) : = \mu_k.$$

Si r_i désigne le rang de E_i / E_{i-1}, on a

$$\mu(E) = \sum_{i=1}^{k} \frac{r_i}{r} \mu_i,$$

donc pour tout sous-faisceau $F \subset E$ on a

$$(*) \quad \mu(F) \leq \mu_{max}(E) \leq \mu(E) + (\mu_{max}(E) - \mu_{min}(E)).$$

<u>REMARQUE</u> :

Si $X = \mathbb{P}_1$ et

$$E = \mathcal{O}(a_1)^{r_1} \oplus \ldots \oplus \mathcal{O}(a_k)^{r_k}, \quad a_1 > \ldots > a_k,$$

alors

$$E_i = \mathcal{O}(a_1)^{r_1} \oplus \ldots \oplus \mathcal{O}(a_i)^{r_i}$$

et $\mu_i = a_i$.

<u>PROPOSITION 2.1.</u>

Soient E et F deux fibrés sur X.

Si

$$\mu_{min}(E) > \mu_{max}(F),$$

alors

$$\mathrm{Hom}(E,F) = 0.$$

<u>DEMONSTRATION</u> :

Soient

$$0 = E_0 \subset E_1 \subset \ldots \subset E_k = E,$$

$$0 = F_0 \subset F_1 \subset \ldots \subset F_\ell = F$$

les filtrations de HARDER-NARASIMHAN de E et F. Un homomorphisme non nul $f : E \longrightarrow F$ induit pour au moins un couple (i,j) un homomorphisme non nul

$$\bar{f} : E_i / E_{i-1} \longrightarrow F_j / F_{j-1}.$$

Comme E_i/E_{i-1} et F_j/F_{j-1} sont semi-stables, on a

$$\mu(E_i/E_{i-1}) \leq \mu(\mathrm{Im}\ \bar{f}) \leq \mu(F_j/F_{j-1}),$$

d'où $\mu_{min}(E) \leq \mu_{max}(F)$.

Introduisons pour un fibré E sur X un certain invariant b(E).

b(E): = sup $\min_i \deg(G_i/G_{i-1})$, où le sup est pris sur toutes les filtrations

$$0 = G_0 \subset G_1 \subset \ldots \subset G_r = E$$

avec G_i/G_{i-1} fibré en droites.

PROPOSITION 2.2.

On a une minoration

$$b(E) \geq \mu_{min}(E) - 2g(r-1)$$

où g est le genre de X et r le rang de E.

DEMONSTRATION.

Soit

$$0 = F_0 \subset F_1 \subset \ldots \subset F_r = E^*$$

une filtration maximale au sens d'ATIYAH [1] . Alors d'après Lemma 4 loc. cit.

$$\deg (F_i/F_{i-1}) \leq \deg (F_1) + 2g(i-1) \leq \mu_{max}(E^*) + 2g(r-1).$$

En dualisant on obtient le résultat.

PROPOSITION 2.3.

Soient E, F, G des fibrés vectoriels sur X. Si

$$\mu_{min}(E) + b(G) > \mu_{max}(F),$$

alors

$$\mathrm{Hom}(G,\mathcal{H}om(E,F)) = 0.$$

DEMONSTRATION.

Soit

$$0 = G_0 \subset G_1 \subset \ldots \subset G_m = G$$

une filtration où tous les G_i/G_{i-1} sont des fibrés en droites avec

$$\deg(G_i/G_{i-1}) \geq b(G).$$

Il suffit de démontrer que

$$\mathrm{Hom}\ (G_i/G_{i-1},\ \mathcal{H}om(E,F)) = 0$$

pour tout i. Comme ce groupe est isomorphe à

$$\mathrm{Hom}\ (E \otimes (G_i/G_{i-1}),F)$$

et

$$\mu_{min}(E \otimes (G_i/G_{i-1})) \geq \mu_{min}(E) + b(G) > \mu_{max}(F),$$ l'assertion

découle de la proposition 2.1.

§ 3. FILTRATION DE HARDER-NARASIMHAN RELATIVE.

Dans ce paragraphe on considère une famille propre et plate $f : X \longrightarrow S$ de courbes lisses, muni d'un fibré $\mathcal{O}_X(1)$ très ample relatif à f. Soit E un fibré vectoriel de rang r sur X.

LEMME 3.1.

Pour $0 < r' < r$ et $\mu' \in \mathbb{Q}$ désignons par $S(r',\mu')$ l'ensemble des points $s \in S$ tels que le fibré E(s) sur la fibre X(s) admette un sous-faisceau F de rang r' avec $\mu(F) = \mu'$.

Alors $S(r',\mu')$ est fermé.

DEMONSTRATION.

S étant irréductible, le polynôme de Hilbert de E(s) est indépendant de s. Par conséquent il existe un polynôme P avec la propriété suivante :

Un sous-faisceau $F \subset E(s)$ est de rang r' avec $\mu(F) = \mu'$ si et seulement si le polynôme de Hilbert de E(s) / F est égal à P.

Soit

$$Q: = \mathrm{Quot}^P_{E/X/S} \xrightarrow{\pi} S$$

la variété projective sur S (non nécessairement irréductible) qui représente les quotients de E(s) de polynôme de Hilbert P (voir GROTHENDIECK [7]). Alors $S(r',\mu')$ est la projection de Q sur S, donc est fermé.

PROPOSITION 3.2.

Il existe une filtration de E par des sous-faisceaux

$$0 = E_0 \subset E_1 \subset \ldots \subset E_k = E$$

telle que sur une fibre générale $X(s)$,

$$0 = E_0(s) \subset E_1(s) \subset \ldots \subset E_k(s) = E(s)$$

soit la filtration de HARDER-NARASIMHAN de $E(s)$.

DEMONSTRATION

On peut supposer S lisse. Avec les notations du lemme précédent soit

$$S(\mu') := \bigcup_{r'} S(r',\mu').$$

Pour $\mu' >> 0$, $S(\mu') = \phi$ et pour $\mu'' << 0$, $S(\mu'') = S$. Il existe donc un μ_1 maximal tel que

$$S(\mu_1) = S.$$

Choisissons r_1 maximal tel que

$$S(r_1,\mu_1) = S.$$

L'ensemble

$$S' := S \setminus \bigcup_{\mu > \mu_1} S(\mu)$$

est un ouvert dense de S. Comme dans la démonstration précédente, on considère le morphisme $\pi : Q \longrightarrow S$ dont $S(r_1,\mu_1)$ est l'image. Sa restriction à S'

est bijective, parce que pour chaque $s \in S'$ le sous-faisceau $F \subset E(s)$ avec $\mu(F) = \mu_1$ et rang $(F) = r_1$ est uniquement déterminé (c'est le sous-fibré semi-stable maximal de $E(s)$). Par conséquent il existe une section

$$\sigma : \quad S' \longrightarrow Q|S'$$

de π . La section σ définit un quotient G de E au dessus de S'. Soit

$$E_1' := \mathrm{Ker}\ (E|X_{S'} \longrightarrow G)$$

et $E_1'' \subset E$ un prolongement de E_1' sur X.

Posons $E_1 := \hat{E}_1''$ (voir § 1). Au dessus d'un ouvert dense $S_1 \subset S'$, $E_1(s)$ est le sous-fibré semi-stable maximal de $E(s)$ et le quotient E/E_1 est localement libre. On recommence la construction avec $(E/E_1)|X_{S_1}$ et ainsi de suite.

§ 4. CONSTRUCTION DE SOUS-FAISCEAUX.

Pour la construction de sous-faisceaux nous aurons besoin du lemme de descente suivant :

LEMME 4.1.

Soit $p : \tilde{Z} \longrightarrow X$ une submersion de variétés lisses à fibres connexes admettant des sections locales. Soit E un fibré vectoriel sur X et $\tilde{F} \subset p^*E$ un sous-fibré.

Si

$$\mathrm{Hom}\,(T_{\tilde{Z}/X}, \mathcal{H}om(\tilde{F}, p^*E/\tilde{F})) = 0,$$

il existe un sous-fibré $F \subset E$ tel que $p^*F = \tilde{F}$.

Ici $T_{\tilde{Z}/X}$ désigne le fibré tangent relatif.

DEMONSTRATION :

Le sous-fibré $\tilde{F} \subset p^*E$ définit un X-morphisme

$$\begin{array}{ccc} \tilde{Z} & \xrightarrow{f} & Gr(E) \\ & {\scriptstyle p}\searrow \quad \swarrow{\scriptstyle \pi} & \\ & X & \end{array}$$

où $Gr(E)$ est la Grassmannienne des k-plans dans les fibres de E, $k = \mathrm{rang}\ \tilde{F}$.

Considérons la différentielle relative de f

$$d_{\tilde{Z}/X}f : T_{\tilde{Z}/X} \longrightarrow f^*\, T_{Gr(E)/X}.$$

On a $T_{Gr(E)/X} = \mathcal{H}om(\Phi, \pi^*E/\Phi)$, où Φ est le sous-fibré tautologique de rang k de π^*E. Donc

$$f^*\, T_{Gr(E)/X} \cong \mathcal{H}om\,(\tilde{F}, p^*E/\tilde{F}).$$

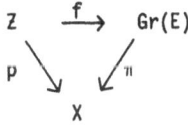

L'hypothèse entraine $d_{Z/X}f = 0$. Par conséquent f est constante sur les fibres de p et induit une application

$$f_0 : X \longrightarrow Gr(E)$$

avec $f = f_0 \circ p$ et $\pi \circ f_0 = id_X$. Comme p admet des sections locales, f_0 est un morphisme algébrique et définit donc un sous-fibré $F \subset E$ avec $p^*F = \tilde{F}$.

La construction standard.

Pour tout le reste du paragraphe, soit X une variété projective lisse de dimension n, munie d'un fibré en droites très ample $\mathcal{O}_X(1)$. Ce fibré définit un plongement

$$X \hookrightarrow \mathbb{P}(V), \quad V := \Gamma(X, \mathcal{O}_X(1))^*,$$

où $\mathbb{P}(V)$ est l'espace projectif des droites de V. Désignons par Gr la Grassmannienne des sous-espaces linéaires de codimension n-1 de $\mathbb{P}(V)$. Pour chaque $Y \in Gr$ l'intersection $X \cap Y$ est non vide et de dimension ≥ 1.
Il existe un ouvert dense $U \subset Gr$ tel que pour tout $Y \in U$, les variétés X et Y se coupent transversalement suivant une courbe lisse irréductible (Bertini).
Désignons par $Drap \subset \mathbb{P}(V) \times Gr$ la sous-variété définie par la relation d'incidence $x \in Y$. On a des projections naturelles

$$\begin{array}{ccc} Drap & \xrightarrow{\bar{q}} & Gr \\ {\scriptstyle \bar{p}}\downarrow & & \\ \mathbb{P}(V) & & \end{array}$$

Les deux projections définissent sur Drap deux structures d'espace fibré (localement trivial). La fibre de $\bar{q}$ est un $\mathbb{P}_{N-n+1}$, $N = \dim \mathbb{P}(V)$, et la fibre de $\bar{p}$ est isomorphe à la Grassmannienne des sous-espaces vectoriels de dimension N-n+1 dans $\mathbb{C}^N$.

Soit $Z \subset$ Drap la sous-variété $\bar{p}^{-1}(X)$ et désignons par p et q les restrictions de $\bar{p}$ et $\bar{q}$ à Z. Alors on a un diagramme

$$\begin{array}{ccc} Z & \xrightarrow{q} & Gr \\ {\scriptstyle p}\downarrow & & \\ X & & \end{array}$$

Il existe un ouvert dense $U \subset Gr$ tel que pour tout $g \in U$, la fibre $q^{-1}(g)$ soit une courbe lisse irréductible C que p plonge dans X.

Soit d_o le plus petit entier tel que pour une fibre générale $C = q^{-1}(g)$, (g parcourant un ouvert dense $U' \subset U$) on ait

$$b(T_{Z/X} \,|C) \geq - d_o, \quad (\text{cf. } \S\ 2).$$

REMARQUE :

Pour $X = \mathbb{P}_n$ les courbes C sont des droites projectives et on a

$$T_{Z/X} \mid C \cong \mathcal{O}_C\ (-1)^{n-1}$$

Donc dans ce cas on trouve $d_o = 1$.

Nous pouvons maintenant énoncer une généralisation du théorème de SPINDLER [15] .

THEOREME 4.2.

Soit E un faisceau sans torsion sur X . On suppose qu'il existe i avec

$$\mu_i(E|C) - \mu_{i+1}\ (E|C) > d_o$$

pour une courbe générale $C = pq^{-1}(g)$.

Alors il existe un sous-faisceau $F \subset E$ tel que pour C générale $F|C$ soit le i-ième terme de la filtration de HARDER-NARASIMHAN de $E|C$.

DEMONSTRATION.

Soit $U \subset Gr$ un ouvert dense tel que $p^* E \mid q^{-1}(U)$ soit un fibré.

Soit

$$0 = E_o \subset E_1 \subset \ldots \subset E_k = p^* E|q^{-1}(U)$$

la filtration de Harder-Narasimhan relative. On peut supposer que les E_i sont des sous-fibrés et que la filtration induit la filtration de Harder-Narasimhan de $E|C$ pour tout $C = pq^{-1}(g)$, $g \in U$.

Appliquons le lemme 4.1. au morphisme

$$p : Z' \longrightarrow X'$$

où $Z' := q^{-1}(U)$, $X' := p(Z')$, et au sous-fibré $E_i \subset p^* E$. Il faut vérifier l'hypothèse

$$\mathrm{Hom}\,(T_{Z'/X'}, \mathcal{H}om(E_i, p^* E/E_i)) = 0.$$

Comme

$$b(T_{Z'/X'} \mid C) \geq - d_o,$$

$$\mu_{min}(E_i|C) = \mu_i(E|C),$$

$$\mu_{max}((p^* E/E_i)|C) = \mu_{i+1}(E|C),$$

l'annulation du groupe en question découle de la proposition 2.3. Le lemme 4.1. nous fournit donc un sous-fibré $F' \subset E|X'$ avec $p^* F'|Z' = E_i$.

Soit $F \subset E$ un prolongement de F'.

C'est le faisceau cherché.

COROLLAIRE.

Soit E un faisceau semi-stable sur X. Alors

$$\mu_i(E|C) - \mu_{i+1}(E|C) \leq d_o$$

pour la courbe générale $C = pq^{-1}(g)$ et tout i.

Pour $X = \mathbb{P}_n$, on retrouve le théorème de SPINDLER [15] .

Nous dirons qu'un faisceau semi-stable E sur X est e-semi-stable (cf. MARUYAMA [13]) si pour la courbe générale $C = pq^{-1}(g)$ tout sous-faisceau non nul $F \subset E|C$ vérifie

$$\mu(F) \leq \mu(E) + e.$$

En utilisant la formule (*) du § 2, on obtient :

<u>COROLLAIRE</u>.

Tout faisceau semi-stable de rang r sur X est e-semi-stable pour $e = (r-1)\, d_o$.

D'après MARUYAMA ([12] , Lemma 3.3) ceci implique que la famille des faisceaux semi-stables sur X ayant un polynôme de Hilbert fixé est limitée. Ce résultat a des conséquences pour les espaces des modules. Par exemple l'espace des modules de faisceaux semi-stables au sens de Gieseker-Maruyama est projectif ([13] , Corollary 5.9.1.).

BIBLIOGRAPHIE

[1] ATIYAH, M.F. Vector bundles over an elliptic curve. Proc. London Math. Soc. (3) 7, 414-452 (1957).

[2] BARTH, W. Some properties of stable rank-2 bundles on P_n, Math. Ann. 226, 125-150 (1977).

[3] ELENCWAJG, G. Les fibrés uniformes de rang 3 sur $P_2(C)$ sont homogènes. Math. Ann. 231, 217-227 (1978).

[4] ELENCWAJG, G. FORSTER, O. Bounding cohomology groups of vector bundles on P_n. Math. Ann. 246. 251-270 (1980).

[5] GIESEKER, D. On the moduli of vector bundles on an algebraic surface, Ann. of Math. 106, 45-60 (1977).

[6] GRAUERT, H. MÜLICH, G. Vektorbündel vom Rang 2 über dem n-dimensionalen komplex-projektiven Raum, Manuscripta math. 16,75-100 (1975).

[7] GROTHENDIECK, A. Techniques de construction et théorèmes d'existence en géométrie algebrique, IV : Les schémas de Hilbert Sém. Bourbaki, 13 (1960/61), exposé 221.

[8] HARDER, G. NARASIMHAN, M.S. On the cohomology groups of moduli spaces of vector bundles on curves, Math. Ann. 212, 215-248 (1975).

[9] LANGTON, S.G. Valuative criteria for families of vector bundles on algebraic varieties, Ann. of Math. 101, 88-110 (1975).

[10] MARUYAMA, M. Stable vector bundles on an algebraic surface, Nagoya Math. J. 58, 25-68 (1975).

[11] MARUYAMA, M. Openness of a family of torsion free sheaves J. Math. Kyoto Univ. 16, 627-637 (1976).

[12] MARUYAMA, M. Moduli of stable sheaves I, J. Math. Kyoto Univ. 17 91-126 (1977).

[13] MARUYAMA, M. Moduli of stable sheaves II, J. Math. Kyoto Univ. 18, 557-614 (1978).

[14] MARUYAMA, M. Boundedness of semi-stable sheaves of small ranks, Preprint 1978.

[15] SPINDLER, H. Der Satz von Grauert-Mülich für beliebige semistabile holomorphe Vektorbündel über dem n-dimensionalen komplex-projektiven Raum. Math. Ann. 243 131-141 (1979).

[16] TAKEMOTO, F. Stable vector bundles on algebraic surfaces, Nagoya Math. J. 47, 29-48 (1972).

[17] VAN DE VEN, A. On uniform vector bundles, Math. Ann. 195, 245-248 (1972).

Mathematisches Institut der Universität,
Roxeler Str. 64, D-4400 MÜNSTER

Institut de Mathématiques, Université de Nice
Parc Valrose, F- 06034 NICE CEDEX

Mathematisches Institut der Universität,
Bunsenstr. 3-5, D- 3400 GÖTTINGEN.

On the classification of algebraic space curves

Robin HARTSHORNE**

§ 1. INTRODUCTION.

In this paper we give a survey of some recent results and open problems concerning the classification of algebraic curves in projective 3-space $\mathbb{P}^3_k$ over an algebraically closed field k.

Why study curves particularly in $\mathbb{P}^3$? One reason is that any abstract nonsingular curve X can be embedded in some projective space $\mathbb{P}^n$, and then by successive general projection can be mapped to $\mathbb{P}^3$ so that the image is still nonsingular. In other words, X can be embedded in $\mathbb{P}^3$. This is not true for $\mathbb{P}^2$: the curves which can be embedded as nonsingular curves in $\mathbb{P}^2$ are special. Of course the degree of the image curve in $\mathbb{P}^3$ is not uniquely determined. There exist embeddings of a given abstract curve X of any sufficiently high degree. Nevertheless, if we classify all nonsingular curves in $\mathbb{P}^3$, we will have included all possible abstract curves.

A second reason for studying curves in $\mathbb{P}^3$ is their close connection with vector bundles. Serre [24] first pointed out the connection between rank 2 vector bundles on a 3-dimensional space and curves in that space. A global section of the vector bundle vanishes on a codimension 2 subset of the space, i.e., a curve, and under certain conditions one can recover the vector bundle from the curve. Now this connection has been generalized to make curves in $\mathbb{P}^3$ correspond to rank r vector bundles on $\mathbb{P}^3$ and also to rank 2 torsion-free coherent sheaves. Recent work of Barth [19] and Gruson and Peskine [22] even suggests that certain curves in $\mathbb{P}^3$ may be useful in constructing rank 2 vector bundles on $\mathbb{P}^4$.

Thus there are several modern reasons for studying curves in $\mathbb{P}^3$ besides the old one that it is a fairly accessible yet nontrivial classification problem. Already in the late 19th century many papers were written describing special classes

---

* Partially supported by NSF Grant MCS 77-03719.

* Department of Mathematics, University of California BERKELEY CA 94720 - U.S.A.

of space curves, culminating in the great treatises of Halphen [3] and Noether [13] which provided a classification of curves up to degree 20. In spite of all these specific data and many general results, one cannot yet say that the classification theory is in a satisfactory state.

Let us examine the problem more closely, and restrict our attention for the moment to irreducible nonsingular curves although there are also many interesting questions to pose concerning reducible and singular curves). We wish to classify all irreducible nonsingular curves Y in $\mathbb{P}^3$. As with most classification questions in algebraic geometry, there is a discrete part and a continuous part of the problem. A curve Y has a degree d determined by its embedding in $\mathbb{P}^3$: this is the number of points in which a general plane meets Y. Then it has a genus g, which is its genus as an abstract curve, defined for example as the dimension of the vector space $H^0(Y,\Omega_{Y/k})$ of everywhere regular differential forms on Y. These numerical invariants d, g divide the set of all curves into classes, so that one can formulate the classification problem in two parts as follows.

PROBLEM 1.1.

For which pairs of integers (d,g) does there exist an irreducible nonsingular curve Y of degree d and genus g in $\mathbb{P}^3$?

PROBLEM 1.2.

For given d, g describe the set of all curves Y of degree d and genus g in $\mathbb{P}^3$ by finding the irreducible components and the dimension of the parameter space which parametrizes the algebraic families of all such curves.

The extent of our ignorance in this subject is illustrated by the fact that the answer to problem (1.1) is not yet known. Halphen gave an answer but his proof rests on subtle general position arguments which are not acceptable today without further justification. So the problem remains open. We will describe the present state of knowledge of this problem in § 2 below.

Embedded in problem (1.2) is a theoretical question, which the ancients took for granted, namely that it makes sense to talk about algebraic families of curves in $\mathbb{P}^3$, and that there exists a universal parameter variety which parametrizes an algebraic family containing each curve exactly once. This theoretical question has a satisfactory modern answer, saying that indeed the set of nonsingular curves of given degree and genus is parametrized by a finite union of finite-dimensional quasi-projective algebraic varieties over k. There are two ways of proving this, using either the Chow variety or the Hilbert scheme.

The Chow variety [23, Ch I, § 9.5.] parametrizes cycles, which are formal linear combinations with nonnegative integer coefficients of irreducible curves in $\mathbb{P}^3$. For cycles of a given degree, the Chow variety is a finite union of projective varieties over k. The nonsingular curves form an open subset of this, considered as cycles with coefficient 1, hence form a finite union of quasiprojective varieties. In particular, it follows that there are only finitely many possible values of the genus g for curves of a given degree d.

The Hilbert scheme [21, exposé 221] parametrizes closed subschemes of $\mathbb{P}^3$ with a given Hilbert polynomial. In the case of curves, the Hilbert polynomial depends only on the degree and the arithmetic genus [7, Ch I, § 7] . So for given degree d and arithmetic genus p_a, the Hilbert scheme is a projective scheme over k parametrizing all 1-dimensional closed subschemes of $\mathbb{P}^3$ with the given d and p_a. The nonsingular curves of degree d and genus p_a appear as an open subset. An old result of mine [5] says that the Hilbert scheme for fixed d and p_a is connected. This does not imply that the parameter space of nonsingular curves of given d ang g is connected, for in proving that result one passes through points corresponding to highly nonreduced schemes.

In any case, the existence part of problem (1.2) is settled, and we are left with the task of finding the irreducible components of the Hilbert scheme of curves of given degree d and genus g. Easy examples show that this space is not irreducible. For instance there are two distinct irreducible families of curves of degree 9 and genus 10 [7, IV, 6.4.3] . A more subtle example of Mumford [12] shows that the Hilbert scheme of nonsingular curves may fail to be reduced, even at the generic point of an irreducible component. Halphen tried to define additional numerical invariants which should separate out the irreducible components of the parameter space, but he did not succeed. So one cannot expect an easy answer to problem (1.2).

In this article we will describe what is known about problem (1.1) -- what curves are known to exist, how to construct them, and what general results are known limiting the possible values of d and g. Then we will discuss the closely related problem of finding the maximum possible genus g of a curve of degree d which is not contained in any surface of degree < k, for a given integer k. In § 4 we discuss the equivalence relation of liaison on space curves and characterize the equivalence classes for this equivalence relation. Finally we will discuss briefly the connection between curves and vector bundles which gives a dictionary between statements involving curves in $\mathbb{P}^3$ and reflexive rank 2 sheaves on $\mathbb{P}^3$.

§ 2. CURVES ON SURFACES OF LOW DEGREE.

One can begin classifying curves in $\mathbb{P}^3$ by studying curves which lie on surfaces of small degree. For surfaces of degrees 1, 2, 3, one can describe quite explicitly what types of curves lie on the surface, and in particular we will see what values of d, g it is possible to obtain in this way. In this section all curves will be irreducible and nonsingular.

If a curve Y of degree d lies in a plane $\mathbb{P}^2$, then its genus is uniquely determined by d, and is given by

$$g = \frac{1}{2}(d-1)(d-2) .$$

An easy way to see this is to use the adjunction formula for Y on $\mathbb{P}^2$, which shows that $\omega_Y \cong \mathcal{O}_Y(d - 3)$, where ω_Y is the canonical sheaf. Then use the fact that $\deg \omega_Y = 2g - 2$, which follows from the Riemann-Roch theorem on Y, to obtain the formula for g. It is also easy to see that for every $d > 0$, there exist irreducible nonsingular curves of degree d in $\mathbb{P}^2$. So all these values of d and g occur.

If a curve Y lies on a nonsingular quadric surface Q, one uses the fact that $Q \cong \mathbb{P}^1 \times \mathbb{P}^1$ by the Segre embedding of $\mathbb{P}^1 \times \mathbb{P}^1$ in $\mathbb{P}^3$. Thus Q has two families of lines on it, and the intersection number of Y with lines of each family defines a bidegree (a,b) for the curve Y. Another way of saying this is that the Picard group of Q, Pic Q, which is the group of linear equivalence classes of divisors on Q, is isomorphic to $\mathbb{Z} \oplus \mathbb{Z}$. Under this isomorphism, the class of Y corresponds to the pair of integers a, b. In terms of a and b it is not hard to compute the degree and genus of Y (using the adjunction formula again, for example). One finds

$$d = a + b$$

$$g = (a - 1)(b - 1) .$$

An irreducible curve Y on Q other than the lines with bidegree (1,0) and (0,1) must have a bidegree (a,b) with $a, b > 0$. Conversely, for any $a, b > 0$, the linear system of curves of that bidegree is very ample, so from Bertini's theorem one concludes that there exist irreducible nonsingular curves Y of the given bidegree (a,b) on Q.

Therefore there are curves in $\mathbb{P}^3$ lying on quadric surfaces with all possible values of d, g given by the formulas above for $a, b > 0$. Note in particular that for given d, several values of g are possible, but they are scattered, nonconsecutive integers in general. It is interesting to find the maximum possible value of g for given d. It occurs for $a = b$ if d is even, or $a = b + 1$ if d is odd. So we find that for curves Y of degree d on a nonsingular quadric surface Q,

$$g \leq \begin{cases} \frac{1}{4} d^2 - d + 1 & \text{if } d \text{ is even} \\ \frac{1}{4} (d^2 - 1) - d + 1 & \text{if } d \text{ is odd.} \end{cases}$$

An analysis of curves on singular quadric surfaces, the quadric cones, shows that the possible values of d, g obtained there are a subset of those obtained on nonsingular quadric surfaces.

Note that the possible values of g for curves of given degree d on a quadric surface are thus all less than the genus of a plane curve of the same degree. This is true more generally for all space curves not contained in a plane, according to the following theorem of Castelnuovo.

THEOREM 2.1.

[7,IV,6.4] If Y is a curve of degree d in $\mathbb{P}^3$, not contained in any plane, then its genus g satisfies

$$g \leq \begin{cases} \frac{1}{4} d^2 - d + 1 & \text{if } d \text{ is even} \\ \frac{1}{4} (d^2 - 1) - d + 1 & \text{if } d \text{ is odd.} \end{cases}$$

Furthermore, (1) for every $d \geq 3$ there exist curves in $\mathbb{P}^3$, not contained in any plane, giving equality in these expressions, and (2) any curve Y for which equality holds must lie on a quadric surface.

This theorem, coupled with obvious inequalities $d > 0$, $g \geq 0$, and the known genus of plane curves, already limits considerably the possible values of d, g for space curves. This is also the starting point of Halphen's work on the classification problem, so now let us state his answer to problem (1.1). Since his proof is doubtful, we state it in the form of a conjecture.

CONJECTURE 2.2.

(Halphen) The possible values of the degree d and the genus g of irreducible nonsingular curves Y in $\mathbb{P}^3$ are determined as follows.

(a) Plane curves, for any $d > 0$, with

$$g = \frac{1}{2}(d-1)(d-2) .$$

(b) Curves on quadric surfaces, for any $a, b > 0$, with

$$d = a + b$$
$$g = (a-1)(b-1) .$$

(c) If Y does not lie on a plane or a quadric surface, then

$$g \leq \frac{1}{6} d(d-3) + 1 .$$

(d) For given $d > 0$, every value of $g \geq 0$ allowed by the bound in (c) is possible.

We have already seen that (a) and (b) are true, and that such curves exist. Statement (c) is also true, and is a special case of a more general theorem we will state in section 3 below. It is the existence statement (d) which is not yet proven. In fact Halphen claims [5, p. 127] to construct curves for all $d > 0$ and all $0 \leq g \leq \frac{1}{6} d(d-3) + 1$ on <u>cubic</u> surfaces. I can't tell if his cubic surfaces are supposed to be nonsingular, but we will see shortly that this last statment, at least for nonsingular cubic surfaces, is false. There are definitely gaps in the possible genera g of curves of degree $d \geq 10$ on nonsingular cubic surfaces.

So let us turn now to the study of curves on a nonsingular cubic surface X in $\mathbb{P}^3$. The situation here is more complicated, but still manageable. One knows [7, Ch. V, § 4] that the cubic surface X is isomorphic to a projective plane $\mathbb{P}^2$ with six points $P_1,\ldots,P_6$ blown up. Denoting by ℓ the total transform of a line in $\mathbb{P}^2$, and by e_i the exceptional curve corresponding to P_i, $i = 1,\ldots,6$, the Picard group Pic X is a free abelian group of rank 7 generated by ℓ, $e_1,\ldots,e_6$. If Y is any irreducible curve on X, not equal to one of the e_i, then one can write the linear equivalence class of Y as

$$Y \sim a\ell - \sum_{i=1}^{6} b_i e_i$$

with $a > 0$ and $b_i > 0$. Furthemore one can compute the degree and genus of Y in terms of the integers a, b_i :

$$d = 3a - \sum b_i$$
$$g = \frac{1}{2}(a-1)(a-2) - \frac{1}{2}\sum b_i(b_i - 1) .$$

The question of which divisor classes contain an irreducible nonsingular curve is more subtle in this case, but can be solved by characterizing the very ample divisors not only on X, but also on the rational surfaces obtained by blowing up fewer than 6 points on $\mathbb{P}^2$. The result is this.

PROPOSITION 2.3.

[7, V, Ex. 4.8] For $d \geq 3$, a divisor class $D = a\ell - \Sigma b_i e_i$ of degree $d = 3a - \Sigma b_i$ on the nonsingular cubic surface X contains an irreducible nonsingular curve Y if and only if the following conditions are satisfied (where we assume $b_1 \geq \ldots \geq b_6$ for simplicity) :

(1) $a > 0$

(2) $b_i \geq 0$

(3) $a \geq b_1 + b_2$

(4) $2a \geq b_1 + b_2 + \dots + b_5$

(5) $a^2 > \sum b_i^2$.

This proposition gives in principle a method for determining all possible values of d, g for curves on a nonsingular cubic surface X. Fix $d = 3a - \sum b_i$. Then there is only a finite number of choices of a, $b_1, \dots, b_6$ for the given d satisfying the inequalities of (2.3). Each such choice determines a value of g by the formula given earlier, and thus we obtain all possible values of g. Unfortunately it is not clear how to express the result in a simple form. Here are some partial results about the set of possible values of g on the cubic surface S.

1. The maximum value of g is that given by Halphen, namely

$$g \leq \frac{1}{6} d(d - 3) + 1 ,$$

and this is attained for all $d \equiv 0 \pmod 3$. If $d \equiv 1,2 \pmod 3$, the maximum value of g attained is the above, less $\frac{2}{3}$ [7, V, Ex. 4.7].

2. Aside from g = 0, there is a <u>minimum</u> value of g, namely

$$g \geq \frac{1}{2} (d - 7) .$$

So for example if d = 10, the value g = 1 is impossible, and there are gaps in the possible values of g. A simple proof of this can be given using an old result of mine [6] which says that if Y is a curve of genus $g > 0$ on a rational surface, then the self-intersection $Y^2 \leq 4g + 5$. The adjunction formula for Y on X says $2g - 2 = Y.(Y + K)$ where K is the canonical divisor. But on the cubic surface $K = -H$, where H is a hyperplane section. Therefore $2g - 2 = Y^2 - d$. Combining with the above inequality for Y^2 gives $2g - 2 + d \leq 4g + 5$ hence $g \geq \frac{1}{2}(d - 7)$. See [7, V, Ex. 4.9] for a more precise statement.

3. Even between the minimum and the maximum above there may be further gaps. Thus for d = 28, for example, the possible values of g are 0, 11-13, 20-25, 27, 29-47, and probably all values from 47 up to the maximum of 117.

4. I guess that all values of g should occur between approximately $\frac{1}{3} d^{3/2}$ and the maximum, but don't know how to prove any such result.

Now what can we say about problem (1.1) ? In the conjecture of Halphen it is the existence part (d) which remains open. It seems likely that on the cubic surface all values of g will be possible from the maximum $\frac{1}{6} d(d-3) + 1$ down to some other bound, as yet unspecified. But below that there are gaps, and one will have to look elsewhere for curves of lower genus.

There is another method of constructing curves in $\mathbb{P}^3$ which works well when the degree is large with respect to the genus. Let Y be an abstract curve of genus g, and look for a very ample divisor D of degree d on Y. If this exists, it will give an embedding of Y in some projective space, which can then be projected to $\mathbb{P}^3$. If we ask that D be a nonspecial divisor, then it is not hard to give a criterion for the existence of D [7, IV, 6.1] and thus one can show that there exist nonsingular curves $Y \subseteq \mathbb{P}^3$ for all $g \geq 2$ and $d \geq g + 3$ with the divisor of hyperplane sections being nonspecial. However, existence questions for very ample special divisors on curves are very difficult, and no satisfactory answer is known in that case.

This method gives curves of low genus which cover many of the gaps left by the curves on cubic surfaces. Even so, gaps remain. For example, if d = 28 and g = 26 or 28 then neither method works. In these two particular cases one can still prove by ad hoc methods that curves of this degree and genus exist. If g = 26, take a curve with d = 14, g = 26 which exists on a cubic surface (it it is the maximum genus for d = 14), and then give it a new embedding in $\mathbb{P}^3$ using the linear system 2 x hyperplane section on the curve. If g = 28,

Joe Harris showed me a construction by taking a plane curve Y of degree 9 and mapping it to $\mathbb{P}^3$ by the linear system of quartic curves with 11 basepoints, 8 of which lie on Y.

For a general solution of Halphen's conjecture, which seems likely to be true, I don't know what methods will work best. One could study curves on surfaces of higher degree. If the surface is general, then a theorem of Max Noether [20, exposé XIX] shows that the only curves on it are complete intersections, and these give very few values of d and g. However, it may be possible to get useful results by studying curves on certain special surfaces, such as the Fermat surface of degree d given by $x_0^d + x_1^d + x_2^d + x_3^d = 0$, or rational surfaces obtained from $\mathbb{P}^2$ by suitable linear systems of curves in $\mathbb{P}^2$.

REFERENCES FOR § 2.

Most of the results of this section appear in my book [7] . See especially IV, § 6 for elementary results on the classification of curves in $\mathbb{P}^3$, and V, § 4 for curves on the cubic surface. These questions are also discussed in more detail in the lecture notes [8] .

For older references there are of course the treatises of Halphen [3] and Noether [13] , and also the survey article of Rohn and Berzolari [17] which gives many other references.

§ 3. MAXIMUM GENUS OF SPACE CURVES.

From the examples we have seen, it seems that curves of higher genus (for fixed degree d) should lie on surfaces of lower degree. The plane curves have genus $\sim \frac{1}{2} d^2$. Curves not lying in a plane, according to Castelnuovo's theorem (2.1) have genus at most $\sim \frac{1}{4} d^2$. Curves not lying on a quadric surface, according to Halphen, have genus at most $\sim \frac{1}{6} d^2$. The general principle is that for fixed d, the maximum genus of curves not lying on surfaces of degree < k is a decreasing function of k. Clearly Halphen believed in this principle. It has not been proved, however, so we will formulate the following problem [10, problem 12] .

PROBLEM 3.1.

For each d, $k > 0$, determine the maximum possible genus g of a nonsingular curve Y in $\mathbb{P}^3$, of degree d and genus g, which does not lie on any surface of degree $< k$.

For $k = 1$ we are asking simply for the maximum genus of all space curves of degree d. By comparing any curve with its general projection into a plane, we see that the genus of a space curve is bounded by the genus of a plane curve of the same degree, so $g \leq \frac{1}{2}(d-1)(d-2)$. This maximum is attained, for every $d > 0$, by a nonsingular plane curve, so the problem is completely answered for $k = 1$.

For $k = 2$ the problem is solved by Castelnuovo's theorem (2.1), and the maximum is attained for all $d \geq 3$. Note that for $d = 1,2$, every curve of degree d is contained in a plane, so the problem is undefined.

For larger values of k, we are looking for a suitable generalization of Castelnuovo's theorem. A complete solution of this problem is not yet known, so we will describe here as much as is known. One finds quite quickly that the nature of the problem differs according to the relative size of d and k. The most satisfactory results are in the case when d is large with respect to k.

THEOREM 3.2.

Assume $d > k(k-1)$. Then for any curve Y satisfying the conditions of (3.1),

$$g \leq \frac{d^2}{2k} + \frac{1}{2}d(k-4) + 1 - \varepsilon ,$$

where $\varepsilon \geq 0$ is a quantity depending on the congruence of d (mod k), namely

$$\varepsilon = \frac{1}{2}f(k - f - 1 + \frac{f}{k}) ,$$

where $d \equiv f \pmod{k}$ and $0 \leq f < k$. Furthermore, for each such d, k, the maximum is attained by a curve lying on a surface of degree k.

This theorem solves problem (3.1) completely in the case $d > k(k-1)$. The statement can be found in Halphen. Modern proofs have been given by Harris [4] (incomplete) and Gruson and Peskine [2]. Let me make a few comments on this theorem.

First of all, how does one arrive at the statement ? We can guess that the curves of maximum genus, not lying on a surface of degree $< k$, will actually lie on a surface of degree k. Next, we can guess that on a given surface, the complete intersection curves have maximum genus. So let Y be the complete intersection of surfaces of degrees k, r. We must assume that $r \geq k$ since Y should not be contained in a surface of degree $< k$. Then one can calculate the degree and genus of Y :

$$d = kr$$

$$g = \frac{1}{2} kr(k + r - 4) + 1.$$

If we eliminate r from these two equations, we obtain exactly the bound of the theorem, with $\varepsilon = 0$. Also, of course, $d \geq k^2$. This gives a heuristic derivation of the formula.

Secondly, there is a simple proof due to Joe Harris that a curve Y lying on a nonsingular surface X of degree k satisfies that inequality (without the ε term). We use intersection theory on X and the Hodge index theorem [7, V, 1.9] in $(\mathrm{Pic}\, X) \otimes \mathbb{Q}$. Let H be a plane section of X. Then $Y.H = d$ and $H^2 = k$. Let $D = Y - \frac{d}{k} H$. Then $D.H = 0$, so by the Hodge index theorem, $D^2 \leq 0$. This gives

$$Y^2 - 2\frac{d}{k} Y.H + \frac{d^2}{k^2} H^2 \leq 0.$$

Substituting for $Y.H$ and H^2, we find $Y^2 \leq \frac{d^2}{k}$. Now the adjunction formula says

$$2g - 2 = Y.(Y + K) .$$

Since the canonical divisor K is (k - 4)H, combining these results gives

$$g \leq \frac{1}{2}\,\frac{d^2}{k} + \frac{1}{2}\,d(k - 4) + 1.$$

Thirdly using methods which are a natural generalization of the proof of Castelnuovo's theorem, one can show in fact that any curve Y, contained in a surface of degree k (possibly singular), but not contained in any surface of lower degree, and with $d > k\,(k - 1)$, satisfies the inequality of (3.2), with ε . A proof of this is given for example in [9, 6.1] . The idea is to intersect the curve with a general plane, and make a careful study of the set of points in the plane.

Even this result does not prove the theorem, however. The point is, that in the absence of a proof of the general principle mentioned above, we cannot assume that the maximum g for curves not contained in surfaces of degree < k is actually attained by a curve which lies on a surface of degree k. The least degree of a surface containing the curve may be some $k' > k$. If $d > k'\,(k' - 1)$, then we can still apply the result [9,6.1] just mentioned. However if k' is too large, some other method is needed. This is the subtle part of the theorem, for which we refer to the paper of Gruson and Peskine [2] .

On the other end of the scale, if d is small with respect to k, elementary estimates using the Riemann-Roch theorem and Clifford's theorem provide a bound for problem (3.1).

THEOREM 3.3.

Let $k \geq 2$. If $d < \frac{1}{6}\,(k^2 + 4k + 6)$ there are no curves of degree d satisfying the conditions of (3.1). If $\frac{1}{6}\,(k^2 + 4k + 6) \leq d < \frac{1}{3}\,(k^2 + 4k + 6)$, then for any curve Y satisfying the conditions of (3.1),

$$g \leq d(k - 1) - \binom{k + 2}{3} + 1.$$

PROOF.

Let $\mathcal{I}_Y$ be the ideal sheaf of Y in $\mathbb{P}^3$. If Y is not contained in any surface of degree < k, then $H^0(\mathbb{P}^3, \mathcal{I}_Y(k-1)) = 0$.

It follows that

$$h^0(Y, \mathcal{O}_Y(k-1)) \geq h^0(\mathbb{P}^3, \mathcal{O}_{\mathbb{P}^3}(k-1)) = \binom{k+2}{3}.$$

If the linear system on Y corresponding to $\mathcal{O}_Y(k-1)$ is special, then Clifford's theorem [7, IV, 5.4] implies that

$$h^0(\mathcal{O}_Y(k-1)) - 1 \leq \frac{1}{2} d(k-1) .$$

Combining with the above inequality gives

$$\frac{1}{2} d(k-1) \geq \binom{k+2}{3} - 1,$$

which is equivalent to $d \geq \frac{1}{3}(k^2 + 4k + 6)$. So this case is outside the hypotheses of the theorem.

We conclude that $\mathcal{O}_Y(k-1)$ is nonspecial. Then by the Riemann-Roch theorem

$$h^0(\mathcal{O}_Y(k-1)) = d(k-1) + 1 - g .$$

Combining with the earlier inequality gives

$$g \leq d(k-1) - \binom{k+2}{3} + 1$$

as required. Since the genus of any irreducible curve is ≥ 0, for Y to exist we must have $g \geq 0$, and this implies $d \geq \frac{1}{6}(k^2 + 4k + 6)$.

This completes the proof.

REMARK 3.3.1.

We do not know whether this bound is the best possible. In other words, for each d, k in the ranges given, does there exist a curve attaining that maximum ?

Even a very special case is nontrivial. Take $d = \frac{1}{6}(k^2 + 4k + 6)$ for suitable k. Then the bound gives g = 0. So the question is, does there exist a rational curve of degree d not contained in any surface of degree < k . Of course there exist rational curves of degree d in $\mathbb{P}^3$ for all $d \geq 1$. Furthermore our hypothesis implies $h^0(\mathcal{O}_{\mathbb{P}^3}(k-1)) = h^0(\mathcal{O}_Y(k-1))$. The hard part is, if the rational curve is sufficiently general, does it follow that the natural map

$$H^0(\mathcal{O}_{\mathbb{P}^3}(k-1)) \longrightarrow H^0(\mathcal{O}_Y(k-1))$$

is injective ? That would imply that Y is not contained in any surface of degree < k.

We now come to the most difficult middle range $\frac{1}{3}(k^2 + 4k + 6) \leq d \leq k(k-1)$, where the solution to problem (3.1) is not yet known. Halphen was aware of the difficulty, and equivocates as follows [3, p.6] : "pour les nombres k, donnant $k^2 - k \geq d$, il n'y a point de formule simple fournissant l'expression de la fonction H(k). Je dois me contenter de dire, dans ce résumé, qu'il existe une méthode certaine pour calculer cette fonction". It is not clear to me how he proposes to calculate this function, even though he does give some examples later on.

Based on examples coming from the theory of vector bundles on $\mathbb{P}^3$ (see § 5 below) I would like to propose the following conjecture.

CONJECTURE 3.4.

Assume $\frac{1}{3}(k^2 + 4k + 6) \leq d \leq k(k-1)$. Then for any curve Y satisfying the conditions of (3.1),

$$g \leq \frac{1}{4}d\,(k - 7 + \sqrt{12d - 3k^2 - 6k + 1}) + 1 \quad .$$

Here are some comments about this conjecture. First of all, I do not expect it to be exact for all values of d and k. There should be an error term, analogous to ε in (3.2), yet to be discovered.

Secondly, a weak evidence in favor of the conjecture is that it dovetails nicely with the results of (3.2) and (3.3). For $d = k^2$ this formula gives the same result as (3.2). For $d = \frac{1}{3}(k^2 + 4k + 6)$, this formula gives a quantity larger by 1 than the expression of (3.3). This is natural, because it will be the first case that $\mathcal{O}_Y(k - 1)$ may be a special divisor, so we expect $h^1 = 1$.

Now let me explain how this formula is derived. There is a conjecture (see § 5) that if E is a stable rank 2 vector bundle on $\mathbb{P}^3$ with $c_1 = 0$, and if $(t+1)(t+3) > 3c_2$, then $H^0(E(t)) \neq 0$. Let us suppose (which is not known) that this is the best possible statement. In other words, let us assume that there exist bundles with $c_1 = 0$ and all $c_2 > 0$ for which $H^0(E(t - 1)) = 0$ when t is the least integer such that $(t + 1)(t + 3) > 3c_2$. Then we take an integer $n \geq t$ and a section $s \in H^0(E(n))$ to obtain a curve Y (see § 5) which will have

$$d = c_2 + n^2$$

$$g = d(n - 2) + 1 \quad .$$

Furthermore, from the exact sequence

$$0 \to \mathcal{O} \to E(n) \to \mathcal{I}_Y(2n) \to 0$$

and the hypothesis $H^0(E(t - 1)) = 0$, we obtain $H^0(\mathcal{I}_Y(n + t - 1)) = 0$. So, taking $k = n + t$, Y will not be contained in any surface of degree k. For n sufficiently large and the section s sufficently general, the curve Y will be irreducible and nonsingular. Also for n sufficiently large, d will lie in the required range with respect to k.

Thus, assuming the above bundles exist, we obtain many curves in the range which interests us. Since t is supposed to be the least integer for which $(t + 1)(t + 3) > 3c_2$, it must satisfy

$$t(t + 2) \leq 3c_2 .$$

Combining this inequality with the equations relating d, g, c_2, n, t, k above yields the expression of the conjecture. (Eliminate n so as to express t and c_2 in terms of d, g, k. Then substitute in the inequality and solve for $(g - 1)/d$ by the quadratic formula).

A final piece of evidence for the conjecture is that it would imply an affirmative solution of the conjecture about $H^0(E(t)) \neq 0$ (see § 5).

§ 4. LIAISON

In this section we consider another approach to the problem of classifying curves in $\mathbb{P}^3$, using the notion of residual intersection. This point of view holds that the complete intersection curves are the most basic. Indeed, if a curve Y is a complete intersection of two surfaces X_1 and X_2 of degrees k_1 and k_2, then it is easy to compute the degree and genus of Y :

$$d = k_1 k_2$$

$$g = \frac{1}{2} k_1 k_2 (k_1 + k_2 - 4) + 1 .$$

Furthermore the totality of such complete intersection curves is described by letting X_1 and X_2 range over all possible surfaces of degrees k_1 and k_2. These are parametrized by projective spaces whose coordinates are the coefficients of the equations of X_1 and X_2. So indeed we may consider that the classification problem for complete intersection curves is solved.

If Y is an arbitrary curve, we may regard it as the <u>partial intersection</u> of two surfaces. Simply take any two surfaces X_1 and X_2 containing Y, and having no common component. The complete intersection of X_1 and X_2 then consists of Y plus some other curve Z which we call the <u>residual intersection</u> of the surfaces

X_1 and X_2 containing Y. There are formulas relating the degrees of the two surfaces X_1 and X_2 with the numerical invariants of the curves Y and Z. Sometimes the residual intersection Z may be simpler than Y. We can recover Y from knowing X_1, X_2 and Z. This provides a method of constructing new curves as residual intersections from known curves. Of course there are nasty technical problems if one wants to be sure that the new curve will be irreducible and nonsingular.

An example of the application of these ideas is in the proof of (3.2) where one shows that the curves for which the maximum is attained are either complete intersections themselves, in the case $d \equiv 0 \pmod{k}$, or are partial intersections with the residual curve being a plane curve of degree k - f, in case $d \not\equiv 0 \pmod{k}$.

The approach to the classification problem we will consider in this section is to put an equivalence relation on the set of all curves in $\mathbb{P}^3$ and then study the equivalence classes. Roughly speaking, we will say that two curves Y and Z are equivalent if they are residual to each other inside the complete intersection of two surfaces. This generates an equivalence relation for which all complete intersection curves are equivalent to the empty curve. The classification problem for equivalence classes of curves should then be simpler than the classification problem for the curves themselves.

To make this theory work well, it is essential to allow singular curves, reducible curves, and curves with nilpotent elements in their structure sheaves. Otherwise we will be blocked at every step by nasty questions of showing that the new curves obtained as residual intersections have desired special properties. The natural generality for this theory is to consider Cohen-Macaulay curves which are generically local complete intersections. In other words, we consider one-dimensional closed subschemes Y in $\mathbb{P}^3$ without embedded points, and such that Y is a local complete intersection at the generic point of each irreducible component. Two such curves Y and Z are said to be linked (liées) if they have no common components, and their scheme-theoretic union $Y \cup Z$ is a complete intersection curve. Note that if X_1 and X_2 are two surfaces with no common component containing Y, and if Y is equal to the complete intersection of X_1 and X_2 at the generic point of each irreducible component of Y, then the residual intersection Z exists and is another curve of the same type, so that Y and Z are linked.

We say that two curves Y and Z are equivalent for the relation of liaison if there exists a sequence of curves $Y = Y_1, Y_2, \ldots, Y_n = Z$ such that for each i, Y_i and Y_{i+1} are linked as above.

The first main result about liaison is the theorem of Apéry and Gaeta, which says that a nonsingular curve Y is equivalent to a complete intersection if and only if it is projectively normal. This result has been reproved in modern language by Peskine and Szpiro [15] . When dealing with not necessarily nonsingular curves, the condition "projectively normal" is replaced by the notion "projectively Cohen-Macaulay". A scheme Y in $\mathbb{P}^n_k$ is said to be projectively Cohen-Macaulay if its homogeneous coordinate ring $k[x_0,\ldots,x_n]/I_Y$ is a Cohen-Macaulay ring. In the case of the Cohen-Macaulay curves we are studying in $\mathbb{P}^3$, this is equivalent to the statement that for all $n \in \mathbb{Z}$, the natural map

$$H^0(\mathbb{P}^3, \mathcal{O}_{\mathbb{P}^3}(n)) \to H^0(Y, \mathcal{O}_Y(n))$$

is surjective.

Here then is the theorem of Apéry-Gaeta-Peskine-Szpiro.

THEOREM 4.1.

A curve Y in $\mathbb{P}^3$, Cohen-Macaulay and generically local complete intersection as above is equivalent by liaison to a complete intersection curve if and only if Y is projectively Cohen-Macaulay.

If one doesn't mind dealing with the empty curve, one could also say that Y is equivalent to the empty curve if and only if Y is projectively Cohen-Macaulay. In particular, the projectively Cohen-Macaulay curves constitute precisely one equivalence class for the relation of liaison.

This suggests that for other curves, one should somehow measure how much they fail to be projectively Cohen-Macaulay. This can be done as follows. Let I_Y be the ideal sheaf of the curve Y, and let

$$M(Y) = \bigoplus_{n \in \mathbb{Z}} H^1(\mathbb{P}^3, I_Y(n)) .$$

This can be regarded as a graded module over the homogeneous coordinate ring $S = k[x_0, x_1, x_2, x_3]$ of $\mathbb{P}^3_k$. Furthermore, in each degree n, $M(Y)_n$ is a finite-dimensional k-vector space, and for n sufficiently large or sufficiently small, it is 0. Therefore M(Y) is a finite-length graded S-module.

From the exact sequence

$$0 \to I_Y \to \mathcal{O}_{\mathbb{P}^3} \to \mathcal{O}_Y \to 0$$

we obtain an exact sequence

$$H^0(\mathbb{P}^3, \mathcal{O}_{\mathbb{P}^3}(n)) \to H^0(Y, \mathcal{O}_Y(n)) \to H^1(\mathbb{P}^3, I_Y(n)) \to 0 .$$

Thus, according to the criterion given above, Y is projectively Cohen-Macaulay if and only if M(Y) = 0. For other curves, the size of the module M(Y) measures the failure of Y to be projectively Cohen-Macaulay.

The module M(Y) associated to a curve Y behaves well with respect to liaison. It is not hard to show that [16,2.3] that if Y and Z are linked by a complete intersection $X_1 \cap X_2$, then M(Z), as a graded S-module, is obtained from M(Y) by first taking the dual as a k-vector space, which inherits a natural S-module structure, and then shifting the degrees by a suitable integer. We write $M(Z) = M(Y)'(\ell)$ for some $\ell \in \mathbb{Z}$.

It follows that if Y and Z are equivalent by liaison, then M(Z) is isomorphic to a shift of either M(Y) or its dual M(Y)'. In other words, the graded S-module M(Y), up to dualizing and shifting degrees, is an <u>invariant</u> of the liaison equivalence class of Y.

A beautiful theorem of Rao [16] shows that in fact the module M characterizes the liaison equivalence class.

THEOREM 4.2.

Two curves Y and Z in $\mathbb{P}^3$, Cohen-Macaulay and generically local complete intersection as above, are equivalent by liaison if and only if the associated graded S-modules M(Y) and M(Z) become isomorphic after a shift of degrees and possible dualizing. Furthermore, if M is any finite-length graded S-module, then there exists an irreducible nonsingular curve Y in $\mathbb{P}^3$ with M(Y) isomorphic to a shift of M.

It is interesting to note that the proof of Rao's theorem makes use of vector bundles on $\mathbb{P}^3$, this time of rank 3.

This theorem suggests that one should attempt to find some classification theory for finite-length graded S-modules, but to the best of my knowledge, practically nothing is known in this direction.

§ 5. VECTOR BUNDLES AND REFLEXIVE SHEAVES

In this section we will discuss the connection between curves in $\mathbb{P}^3$ and vector bundles on $\mathbb{P}^3$, and see how the theory of vector bundles can illuminate the problem of classifying space curves. The study of vector bundles on projective spaces has been pursued very actively in the last few years, and this is not the place to summarize that work. See for example [9] , [10] , [11] , [14] , [18] , and the references cited therein. Rather we will mention only those results and conjectures which bear most directly on the theory of space curves.

Let E be a rank 2 vector bundle on $\mathbb{P}^3$, and let $s \in H^0(\mathbb{P}^3, E)$ be a global section. Assuming that s does not vanish on any surface in $\mathbb{P}^3$, the zero set Y of s will be a curve in $\mathbb{P}^3$. In fact Y comes with a naturally defined scheme structure which makes it a locally complete intersection curve. The section s and the curve Y are related to E by the exact sequence

$$0 \to \mathcal{O}_{\mathbb{P}^3} \xrightarrow{s} E \to \mathcal{I}_Y(c_1) \to 0,$$

where $c_1 \in \mathbb{Z}$ is the first Chern class of E. The degree and genus of Y are related to the Chern classes c_1 and c_2 of E by the formulas

$$d = c_2$$
$$g = \frac{1}{2} c_2(c_1 - 4) + 1 .$$

Thus a bundle E and a section s give rise to a curve. Replacing the bundle E by a suitable twist $E(n)$, $n \geq 0$, if necessary, and taking the section s sufficiently general, the curve Y will be irreducible and nonsingular.

Conversely, one can recover the bundle E and its section s from the curve Y and certain supplementary data as follows [9,1.1.]. The exact sequence above expresses E as an extension of coherent sheaves of $\mathcal{I}_Y(c_1)$ by $\mathcal{O}_{\mathbb{P}^3}$. This extension determines an element

$$\xi \in \mathrm{Ext}^1_{\mathbb{P}^3}(\mathcal{I}_Y(c_1), \mathcal{O}_{\mathbb{P}^3}) .$$

One can show that this group is isomorphic to $H^0(Y, \omega_Y(4 - c_1))$, where ω_Y is the dualizing sheaf of Y, and that ξ regarded as a section of $\omega_Y(4 - c_1)$ generates that sheaf and so gives an isomorphism $\mathcal{O}_Y \cong \omega_Y(4 - c_1)$.

In this way we obtain a one-to-one correspondence between pairs (E,s) where E is a rank 2 vector bundle on $\mathbb{P}^3$, and s is a global section vanishing in codimension 2, and pairs (Y,ξ), where Y is a local complete intersection curve in $\mathbb{P}^3$, and ξ is a section of $\omega_Y(4 - c_1)$ inducing an isomorphism $\mathcal{O}_Y \xrightarrow{\sim} \omega_Y(4 - c_1)$.

This gives an intimate connection between the theory of rank 2 vector bundles on $\mathbb{P}^3$ and the theory of local complete intersection curves in $\mathbb{P}^3$ for which

$\omega_Y \cong \mathcal{O}_Y(\ell)$ for some $\ell \in \mathbb{Z}$. Note that a (global) complete intersection curve has this property, and the corresponding vector is decomposable, i.e., a direct sum of two line bundles $\mathcal{O}(a) \oplus \mathcal{O}(b)$ on $\mathbb{P}^3$.

Of course a given rank 2 vector bundle E on $\mathbb{P}^3$ may not have any non-zero global sections. In that case, in order to make it correspond to a curve, we must replace it by a suitable twist $E(n)$, $n > 0$, and use a section $s \in H^0(E(n))$. There are many choices for n and s, so the bundle E can be made to correspond to many different curves. However, one can show that for any two such choices n, s and n', s', the corresponding curves Y and Y' are equivalent by liaison [16, 2.7] . The graded modules M(Y) and M(Y') of the curves Y and Y' are isomorphic, up to shift, to the graded S-module

$$M(E) = \bigoplus_{n \in \mathbb{Z}} H^1(\mathbb{P}^3, E(n)) .$$

The converse, however, is not true : there are easy examples* of bundles E , and E', not isomorphic even after a twist, which correspond to curves in the same liaison equivalence class. In other words, the module M(E) does not characterize isomorphism classes of bundles up to twist. Also, one does not know which finite graded S-modules M can occur as M(E) for some rank 2 vector bundle E . Rao [16,3.3] has given examples showing that not all modules M can occur.

Now we make some comments on the classification of rank 2 vector bundles on $\mathbb{P}^3$. The Chern classes c_1, $c_2 \in \mathbb{Z}$ are numerical invariants which are constant in continuous families. They must satisfy $c_1c_2 \equiv 0 \pmod 2$, and all pairs of integers c_1, c_2 satisfying this condition occur as Chern classes of rank 2 vector bundles [9, 2.5] . If one twists a bundle E by an integer n, then $c_1(E(n)) = c_1(E)+2n$.

---

* There is a 5-parameter family of non-isomorphic stable rank 2 bundles on $\mathbb{P}^3$ with $c_1 = 0$, $c_2 = 1$ [9,8.4.1] which all have the same module $M \cong k$.

Therefore it is convenient to <u>normalize</u> E, replacing it by a twist which will have $c_1 = 0$ or -1.

If E is a bundle with $c_1 = 0$ or -1, we define another numerical invariant t to be the least integer such that $H^0(E(t)) \neq 0$. Clearly t is a useful number to know, because one can in principle classify bundles with given c_1, c_2, t, by classifying all curves corresponding to sections of $H^0(E(t))$. Since the degree and genus of these curves is determined by c_1, c_2, t, the curves form a bounded family, parametrized by part of the Hilbert scheme. In particular, the set of bundles with given c_1, c_2, t forms a bounded family.

One can show that for given c_1 and c_2, the invariant t is bounded above [9,8.2]. On the other hand, t is not bounded below, and the set of all bundles with given c_1 and c_2 does not form a bounded family. However, from what was said above, if one fixes t_0 and considers bundles with fixed c_1, c_2, and $t \geq t_0$, these do form a bounded family. Bundles with $c_1 = 0$ or -1 and $t > 0$ are called <u>stable</u> vector bundles. For these bundles one knows that a coarse moduli scheme exists, and considerable effort has gone into the study of the moduli space (see the references mentioned earlier).

Of particular interest in connection with the maximum genus of space curves discussed in § 3 is the precise determination of an upper bound for t as a function of c_1 and c_2. The best possible is not yet known, but one has the following conjecture, which for simplicity I state only in the case $c_1 = 0$.

CONJECTURE 5.1.

Let E be a rank 2 vector bundle on $\mathbb{P}^3$ with $c_1 = 0$, and $c_2 \geq 0$. Let t be an integer such that $t > \sqrt{3c_2+1} - 2$. Then $H^0(E(t)) \neq 0$.

Atiyah pointed out that the conjecture is true for the important class of <u>instanton bundles</u>, which are rank 2 bundles with $c_1 = 0$ and $H^1(E(-2)) = 0$ [9,8.2.3]. On the other hand, the weaker result proved in [9,8.2] shows that the conjecture is true for all $0 \leq c_2 \leq 9$. More recently, using a result of Barth and Elencwajg

[1] in characteristic 0, I was able to show that the conjecture is true for all odd $c_2 \leq 37$. A similar calculation should work for even c_2. The numerical evidence for the conjecture is thus favorable.

Further evidence for this conjecture (perhaps circular ?) is that it is a consequence of conjecture (3.4). Given a bundle E, take n large and take a section $s \in H^0(E(n))$ corresponding to an irreducible nonsingular curve Y. Let t be an integer for which $H^0(E(t)) = 0$. Then $H^0(\mathcal{I}_Y(t + n)) = 0$, so taking $k = n + t + 1$, Y is not contained in any surface of degree $< k$. The degree and genus are as given in the derivation of (3.4), namely

$$d = c_2 + n^2$$

$$g = d(n - 2) + 1 .$$

Using these values of d, g, k, the inequality of (3.4) gives $t^2 + 4t + 3 \leq 3c_2$. This says $t \leq \sqrt{3c_2 + 1} - 2$, so (5.1) is proved.

A closely related question, already alluded to in § 3, is whether the statement of (5.1) is the best possible. I believe it should be, and a sufficiently general instanton bundle should provide an example. The Riemann-Roch theorem for a rank 2 bundle with $c_1 = 0$ says

$$\chi(E(t)) = \frac{1}{3}(t + 1)(t + 2)(t + 3) - c_2(t + 2) .$$

The value of t given in the conjecture is the least positive value of t which makes $\chi(E(t)) > 0$. This suggests the following conjecture.

CONJECTURE 5.2.

For $c_1 = 0$ and any $c_2 > 0$, there exist rank 2 bundles E on $\mathbb{P}^3$ with the following property : for each $n \in \mathbb{Z}$, at most one of the four groups $H^i(\mathbb{P}^3, E(n))$, $i = 0, 1, 2, 3$, is non-zero.

If this property holds, then the dimension of the corresponding group would be determined by the Riemann-Roch theorem. Furthermore, $H^i(E(-2)) = 0$ for all i, since $\chi(E(-2)) = 0$, so E would be an instanton bundle. By semicontinuity, this property would continue to hold on an open subset of the moduli space, so it would hold for all sufficiently general instanton bundles in some irreducible component of the moduli space.

The difficulty with proving a result like this is that, like a "sufficiently general curve of genus g", it is very hard to lay hands on a "sufficiently general instanton bundle". One approach to this difficulty is to start with a special instanton bundle, such as the one corresponding to a set of skew lines in $\mathbb{P}^3$ (see [9, 4.3.1]), and then use deformation theory to discover the properties of a nearby general bundle. In attempting to do this I was blocked by the following problem in projective geometry.

PROBLEM 5.3.

Let Y be a disjoint union of r lines in $\mathbb{P}^3$ and let k be an integer such that $h^0(\mathcal{O}_{\mathbb{P}^3}(k)) \leq h^0(\mathcal{O}_Y(k))$, i.e. $\binom{k+3}{3} \leq r(k+1)$. If the lines are in sufficiently general position, is it true that Y is not contained in any surface of degree k ?

This is a type of general position question, analogous to the one raised in (3.3.1) above, which apparently needs some new technique to answer.

We close this section with some remarks on reflexive sheaves. One defect of the correspondence between bundles and curves described above is that the vector bundles correspond only to the curves Y with $\omega_Y \cong \mathcal{O}_Y(\ell)$ for some ℓ . This is a serious restriction. In fact, it even limits the possible values of the degree and genus of the curve, as the equation $2g - 2 = d\ell$ shows. To remedy this situation, we need some more general objects on $\mathbb{P}^3$ to correspond to the other curves.

A natural class of objects to consider is the rank 2 reflexive coherent sheaves on $\mathbb{P}^3$. A coherent sheaf is said to be reflexive if the natural map into its double dual $F \to F^{\vee\vee}$ is an isomorphism. Here the dual $F^{\vee}$ of any coherent sheaf F is defined as the sheaf $Hom(F, \mathcal{O}_{\mathbb{P}^3})$. Practically everything mentioned above concerning rank 2 vector bundles carries over to reflexive sheaves. Their sections determine curves. The sheaf F can be recovered from the corresponding curve Y and a section $\xi \in H^0(Y, \omega_Y(4 - c_1))$ which need only generate the sheaf $\omega_Y(4 - c_1)$ at the generic points of the irreducible components of Y. The curves which arise in this way consist of all Cohen-Macaulay generically local complete intersection curves, as in § 4. The reflexive sheaves are locally free except at a finite number of points of $\mathbb{P}^3$, and the number of these points, properly counted, is the third Chern class* $c_3(F)$. This number is also equal to the number of points at which the section ξ fails to generate $\omega_Y(4 - c_1)$. The numerical invariants of F and Y are related by the formulas

$$d = c_2$$

$$g = \frac{1}{2} c_2(c_1 - 4) + \frac{1}{2} c_3 + 1 \quad .$$

Conjecture (5.1), or better a version involving c_3 as well, can be stated for reflexive sheaves, and will correspond to conjecture (3.4) for more general curves.

A final motivation for studying reflexive sheaves is that they arise naturally when one tries to prove results about vector bundles using induction on c_2. I hope to make a systematic study of rank 2 reflexive sheaves on $\mathbb{P}^3$ in a future article.

---

* Chern classes are defined on the Grothendieck group $K^{\cdot}(X)$ of vectorbundles on a variety X. If X is nonsingular, then $K^{\cdot}(X)$ is equal to the Grothendieck group $K_{\cdot}(X)$ of coherent sheaves on X, so we can speak of Chern classes of any coherent sheaf.

REFERENCES

[1] BARTH W. and ELENCWAJG G. Concernant la cohomologie des fibrés algébriques stables sur $\mathbb{P}_n(\mathbb{C})$, in Variétés Analytiques Complexes (Nice, 1977), Lecture Notes in Math. 683 Springer (1978). 1-24.

[2] GRUSON, L. and PESKINE, C. Genre des courbes de l'espace projectif, in Algebraic Geometry, Troms ø 1977, Lecture Notes in Math. 687 Springer (1978) 31-59.

[3] HALPHEN, G. Mémoire sur la classification des courbes gauches algébriques, J. Ec. Polyt. 52 (1882) 1-200.

[4] HARRIS, J. The genus of space curves, preprint, Sept 1978.

[5] HARTSHORNE, R. Connectedness of the Hilbert Scheme, Publ. Math. I.H.E.S. 29 (1966) 6-48.

[6] HARTSHORNE, R. Curves with high self-intersection on algebraic surfaces, Publ. Math. I.H.E.S. 36, (1969), 111-125.

[7] HARTSHORNE, R. Algebraic Geometry, Graduate Texts in Math. 52, Springer Verlag, New-York (1977), xvi + 496 pp.

[8] HARTSHORNE, R. The classification of curves in $\mathbb{P}^3$ and related topics (lecture notes in Japanese by M. Ishida and E. Sato) Math. Res. Lec. Notes 2, Kyoto Univ. (1977), 74 pp.

[9] HARTSHORNE, R. Stable vector bundles of rank 2 on $\mathbb{P}^3$, Math. Ann. 238 (1978) 229-280.

[10] HARTSHORNE, R. Algebraic vector bundles on projective spaces : a problem list. Topology 18 (1979) 117-128.

[11] HARTSHORNE, R. Four years of algebraic vector bundles :1975-1979, Journées de géométrie algébrique d'Angers (1979).

[12] MUMFORD, D. Further pathologies in algebraic geometry, Amer. J. Math. 84 (1962), 642-648.

[13] NOETHER, M. Zur Grundlegung der Theorie der algebraischen Raumcurven, Verlag der Königlichen Akademie der Wissenschaften, Berlin (1883).

[14] OKONEK, C. SCHNEIDER, M. SPINDLER, H. Vector bundles on complex projective spaces, Birkhäuser Boston, Inc. (To appear).

[15] PESKINE, C. and SZPIRO, L. Liaison des variétés algébriques (I), Invent. math. 26 (1974) 271-302.

[16] RAO, A.P. Liaison among curves in $\mathbb{P}^3$, Invent. math. 50 (1979) 205-217.

[17] ROHN, K. BERZOLARI, L. Algebraische Raumkurven und abwickelbare Flächen, Encyclop. d. math. Wiss. IIIC9 (1926), 1229-1436.

[18] SCHNEIDER, M. Holomorphic vector bundles on $\mathbb{P}_n$, Séminaire Bourbaki, 530 (1978/79), 23 pp.

[19] BARTH, W. Kummer surfaces associated with the Horrocks-Mumford bundle, Angers Proceedings (1979) preprint.

[20] DELIGNE, P. and KATZ, N. Groupes de monodromie en géométrie algebrique (SGA 7, II), Lecture notes in math. 340, Springer (1973).

[21] GROTHIENDIECK, A. Fondements de la géométrie algébrique (extraits du Séminaire Bourbaki), Secr. Math. Paris (1962).

[22] GRUSON, L. An introduction to a critical reading of Halphen's article, Lectures at Nordic Summer School in Algebraic Geometry, Stockholm (1979), unpublished.

[23] SAMUEL, P. Méthodes d'algèbre abstraite en géométrie algébrique, Ergebnisse der Math. 4 Springer (1955).

[24] SERRE, J.P. Sur les modules projectifs. Séminaire Dubreil-Pisot (1960/61) Secr. Math. Paris, exposé 2.

ON THE CLASSIFICATION OF STABLE RANK-r VECTOR BUNDLES OVER THE PROJECTIVE PLANE

Klaus HULEK

0. INTRODUCTION :

Though quite a lot is known on the classification of stable rank-2 vector bundles over the projective plane, this is not the case for arbitrary rank. In my talk I wanted to give some results concerning the moduli of rank-r vector bundles. We shall restrict ourselves to $c_1(\mathcal{F}) = 0$. The reason is that though most of the ideas should also work for $c_1(\mathcal{F}) \neq 0$ the technical modifications which are necessary, are quite substantial, and so far I have not been able to bring them into a unified theory.

In the first chapter we shall characterize those vector bundles $\mathcal{F}$ with $h^o(\mathcal{F}) = h^o(\mathcal{F}^*) = 0$ (which holds for stable bundles) by means of the map $H^1(\mathcal{F}(-2)) \otimes \Gamma(\mathcal{O}(1)) \longrightarrow H^1(\mathcal{F}(-1))$. This will lead us to equivalence classes of certain "Kronecker-modules" $\alpha : \mathbb{C}^n \otimes \Gamma(\mathcal{O}(1)) \longrightarrow \mathbb{C}^n$.
One of the main points there will be to determine the rank of a bundle $\mathcal{F}$ which belongs to a given α .

In the second chapter we shall make use of the above construction to prove that the moduli-scheme of stable vector bundles with given rank and given second Chern class will be irreducible.

The last chapter is concerned with stability. We first recall the definition of stability given by Mumford-Takemoto.
Then we consider stability (in the sense of Mumford) of the Kronecker-modules under the operation of the group $SL(n,\mathbb{C}) \times SL(n,\mathbb{C})$. We prove that for all rank-2 bundles and that for rank-3 bundles with $c_2(\mathcal{F}) \leq 7$ the bundle $\mathcal{F}$ is stable if and

only if $\alpha_{\mathcal{F}}$ is stable. As an application we get the Grauert-Mülich theorem for rank-2 bundles. Throughout this paper we fix a 3-dimensional vector space V and set $\mathbb{P}_2 = \mathbb{P}_2(V)$. Then we have natural identifications $V^* = \Gamma(\mathcal{O}(1))$ and $V = \Gamma(\Omega(2))$. For the sake of convenience we fix a basis $v_o, v_1, v_2 \in V$ and its dual $z_o, z_1, z_2 \in V^*$. We shall identify $\Lambda^2 V = V^*$ via $v_{i+1} \wedge v_{i-1} \longrightarrow z_i$ and $\Lambda^3 V = \mathbb{C}$ via $v_o \wedge v_1 \wedge v_2 \longrightarrow 1$.

I want to thank W. Barth for many helpful discussions.

1. THE MONAD CONSTRUCTION.

1.1. Throughout this paper we shall consider rank-r vector bundles $\mathcal{F}$ with $c_1(\mathcal{F}) = 0$ and $c_2(\mathcal{F}) = n$. We are interested in *stable* vector bundles (in the sense of Mumford-Takemoto). All what we need of stability for the first two chapters will be that a stable bundle is simple and the bundle as well as its dual does not have sections (c.f. (3.1.3).

1.1.1. **Definition.** A vector bundle $\mathcal{F}$ will be called *s-stable* if $h^o(\mathcal{F}) = h^o(\mathcal{F}^*) = o$.

Remark. Stable bundles are s-stable but the converse is only true if $\text{rank}(\mathcal{F}) \leq 3$ (See (3.1.3.)).

1.1.2. **Lemma.** *If $\mathcal{F}$ is s-stable then*

$$h^1(\mathcal{F}(-2)) = h^1(\mathcal{F}(-1)) = n.$$

Proof. We have $h^o(\mathcal{F}(k)) = h^o(\mathcal{F}^*(k)) = 0$ for $k \leq 0$ which by Serre-duality implies $h^2(\mathcal{F}(k)) = 0$ for $k \geq -3$. Hence for $-3 \leq k \leq 0$

$$h^1(\mathcal{F}(k)) = -\chi(\mathcal{F}(k)) = n - r(1 + \frac{3}{2}k + \frac{1}{2}k^2) .$$

Remark. As we consider bundles which at least have rank 2 it follows immediately from above that $n \geq 2$ and that for a given second Chern class n the rank of the bundle will be bounded by $2 \leq r \leq n$.

1.2. For a given vector bundle $\mathscr{F}$ we want to study the multiplication map

$$\alpha \;:\; H^1(\mathscr{F}(-2)) \otimes V^* \longrightarrow H^1(\mathscr{F}(-1)).$$

For any $z \in V^* = \Gamma(\mathcal{O}(1))$ we define the map

$$\alpha_{\mathscr{F}}(z) := \alpha_{\mathscr{F}}(- \otimes z).$$

Moreover we denote

$$\alpha_i := \alpha_{\mathscr{F},i} := \alpha\,(z_i)\;.$$

It will be the main point of this chapter to describe the bundles $\mathscr{F}$ by means of the map $\alpha_{\mathscr{F}}$.

1.2.1. **Definition.** A homomorphism $\alpha : \mathbb{C}^n \otimes V^* \longrightarrow \mathbb{C}^n$ will be called a *Kronecker-module*.

1.2.2. **Definition.** A Kronecker-module α will be called *pre-stable* if the following holds : For every $0 \neq \varphi \in \mathbb{C}^n$ we have

$$\dim\ \alpha(\varphi \otimes V^*) \geq 2 \text{ and } \dim(\alpha^T(\varphi \otimes V^*)) > 2$$

Remark. (i) α is pre-stable if and only if the subspaces spanned by $\alpha_o(\varphi)$, $\alpha_1(\varphi)$, $\alpha_2(\varphi)$ and $\alpha_o^T(\varphi)$, $\alpha_1^T(\varphi)$, $\alpha_2^T(\varphi)$ respectively are of dimension at least two. In particular $n \geq 2$.

(ii) Another condition equivalent to pre-stability is the following : there are no automorphisms ω_1, $\omega_2 \in GL(n,\mathbb{C})$ such that the matrices $\omega_1^T \alpha_i \omega_2$; $i = 0,1,2$, are all of the form

$$\omega_1^T \alpha_i \omega_2 = \begin{pmatrix} 0 & \\ \vdots & * \\ 0 & \\ * & \end{pmatrix}$$

or of the form

$$\omega_1^T \alpha_i \omega_2 = \begin{pmatrix} 0 \ldots 0 & * \\ & \\ & * \end{pmatrix}.$$

1.2.3. The algebraic group $G = GL(n,\mathbb{C}) \times GL(n,\mathbb{C})$ operates on the set of all Kronecker-modules by

$$\alpha(z) \longmapsto \omega_1 \, \alpha(z) \, \omega_2^T .$$

We shall be interested in the equivalence-classes under this group-operation. The class belonging to α will be denoted by $[\alpha]$.

1.2.4. For every Kronecker-module $\alpha : \mathbb{C}^n \otimes V^* \longrightarrow \mathbb{C}^n$ we define a homomorphism $A(\alpha) : \mathbb{C}^n \otimes V \longrightarrow \mathbb{C}^n \otimes V^*$ by

$$\varphi \otimes v_i \longrightarrow \alpha_{i+1}(\varphi) \otimes z_{i-1} - \alpha_{i-1}(\varphi) \otimes z_{i+1}$$

where $i = 0,1,2 \bmod 3$. If $e_1,\dots,e_n$ is the standard basis of $\mathbb{C}^n$, and if we choose $e_1 \otimes v_o,\dots,e_n \otimes v_o,\dots,e_n \otimes v_2$ as a basis of $\mathbb{C}^n \otimes V$, and similarly $e_1 \otimes z_o,\dots,e_n \otimes z_2$ as a basis of $\mathbb{C}^n \otimes V^*$, then $A(\alpha)$ will be given by

$$A(\alpha) = \begin{pmatrix} 0 & \alpha_2 & -\alpha_1 \\ -\alpha_2 & 0 & \alpha_o \\ \alpha_1 & -\alpha_o & 0 \end{pmatrix}$$

1.3. **Lemma.** *If* $\mathcal{F}$ *is s-stable then the associated Kronecker-module* $\alpha_{\mathcal{F}}$ *will be pre-stable with* rank $A(\alpha_{\mathcal{F}}) = 2n+r$.

Proof. (i) we shall first prove pre-stability. Therefore we assume that there is some $0 \neq \varphi \in H^1(\mathcal{F}-2))$ such that $\alpha(\varphi \otimes V^*)$ has dimension at most one. Then we can find linearly independant elements $z, z' \in V^*$ such that $a(z)\varphi = a(z')\varphi = 0$. Take $p \in \mathbb{P}_2$ to be the intersection of the two lines $\{z = 0\}$ and $\{z' = 0\}$. Then there is an exact sequence

$$0 \longrightarrow \mathcal{F}(-2) \xrightarrow{\binom{z}{z'}} 2\mathcal{F}(-1) \xrightarrow{(-z',z)} \mathcal{I}_p\mathcal{F} \longrightarrow 0$$

from which we get

$$0 \longrightarrow H^o(\mathcal{I}_p.\mathcal{F}) \longrightarrow H^1(\mathcal{F}-2)) \xrightarrow{\binom{\alpha(z)}{\alpha(z')}} 2H^1(\mathcal{F}-1))$$

and hence $h^o(\mathcal{I}_p.\mathcal{F}) \neq 0$ a contradiction to $h^o(\mathcal{F}) = 0$. We can finish the proof by exploiting $h^o(\mathcal{F}^*) = 0$ in the same way.

(ii) The Koszul complex on $\mathbb{P}_2$ is of the form

$$(1)\quad \begin{array}{ccccccccc} 0 \to \mathcal{O}(-3) & \xrightarrow{s} & V\otimes\mathcal{O}(-2) & & \xrightarrow{\wedge s} & & V^*\otimes\mathcal{O}(-1) & \xrightarrow{\wedge s} & \mathcal{O} \to 0 \\ & & & f \searrow & & \nearrow g & & & \\ & & & & \Omega & & & & \\ & & & \nearrow & & \searrow & & & \\ & & & 0 & & 0 & & & \end{array}$$

where s is the tautological bundle map, and where the middle map $\wedge s$ over a point $x \in \mathbb{P}_2$ is given by $v_i \longrightarrow z_{i+1}(x) \otimes z_{i-1} - z_{i-1}(x)\otimes z_{i+1}$.

Tensoring with $\mathcal{F}$ we get

$$(2)\quad \begin{array}{ccccccccc} 0 \longrightarrow \mathcal{F}(-3) & \xrightarrow{id\otimes s} & \mathcal{F}(-2)\otimes V & & \xrightarrow{id\otimes\wedge s} & & \mathcal{F}(-1)\otimes V^* & \xrightarrow{id\otimes\wedge s} & \mathcal{F} \longrightarrow 0 \\ & & & F \searrow & & \nearrow G & & & \\ & & & & \mathcal{F}\otimes\Omega & & & & \\ & & & \nearrow & & \searrow & & & \\ & & & 0 & & 0 & & & \end{array}$$

with $F = id\otimes f$ and $G = id\otimes g$. On the H^1-level this gives

$$\begin{array}{ccccccc} 0 \longrightarrow H^1(\mathcal{F}(-3)) \longrightarrow H^1(\mathcal{F}(-2))\otimes V & & \xrightarrow{A(\alpha_{\mathcal{F}})} & & H^1(\mathcal{F}(-1))\otimes V^* \longrightarrow H^1(\mathcal{F}) \longrightarrow \\ & F \searrow & & \nearrow G & \\ & & H^1(\mathcal{F}\otimes\Omega) & & \end{array}$$

Because of $h^2(\mathcal{F}(-3)) = h^o(\mathcal{F}^*) = 0$ the map F is surjective, whereas $h^o(\mathcal{F}) = 0$ implies that G is injective. Hence

$$\text{rank } (A(\alpha_{\mathcal{F}})) = h^1(\mathcal{F}\otimes\Omega) .$$

But it follows from (2) that $h^o(\mathcal{F}\otimes\Omega) = 0$ because of $h^o(\mathcal{F}(-1)) = 0$.

Similarly one finds $h^o(\mathcal{F}^*\otimes\Omega) = h^2(\mathcal{F}\otimes\Omega) = 0$. Now one can apply Riemann-Roch to find

$$h^1(\mathcal{F}\otimes\Omega) = - \chi(\mathcal{F}\otimes\Omega) = 2n+r.$$

1.4. Our next task will be to associate a vector bundle $\mathcal{F}$ to a given Kronecker-module α .

1.4.1. For this purpose we define maps

$$a: \quad \mathbb{C}^n \otimes \mathcal{O}(-1) \xrightarrow{\mathrm{id}\otimes s} \mathbb{C}^n \otimes V \otimes \mathcal{O} \xrightarrow{A(\alpha)} \mathbb{C}^n \otimes V^* \otimes \mathcal{O}$$

$$c: \quad \mathbb{C}^n \otimes V^* \otimes \mathcal{O} \xrightarrow{\mathrm{id}\otimes\wedge s} \mathbb{C}^n \otimes \mathcal{O}(1) \ .$$

We put $U: = \operatorname{Im} A(\alpha)$ and get a complex

$$M(\alpha) \quad \mathbb{C}^n \otimes \mathcal{O}(-1) \xrightarrow{a} U \otimes \mathcal{O} \xrightarrow{c} \mathbb{C}^n \otimes \mathcal{O}(1) \ .$$

1.4.2. **Lemma.** *If* α *is pre-stable then* $M(\alpha)$ *is a monad and its cohomology bundle* $\mathcal{F}(\alpha)$ *is s-stable with* $c_1(\mathcal{F}) = 0$, $c_2(\mathcal{F}) = n$ *and* $\operatorname{rank} \mathcal{F}(\alpha) = \operatorname{rank} A(\alpha) - 2n$.

Proof. (i) $c \circ a = 0$: Over a point $x \in \mathbb{P}_2$ the map a is given by

$$a(\varphi) = (z_1(x)\alpha_2(\varphi) - z_2(x)\alpha_1(\varphi)) \otimes z_o + (z_2(x)\alpha_o(\varphi) - z_o(x)\alpha_2(\varphi)) \otimes z_1$$
$$+ (z_o(x)\alpha_1(\varphi) - z_1(x)\alpha_o(\varphi)) \otimes z_2$$

and

$$c \circ a(\varphi) = z_o(x)(z_1(x)\alpha_2(\varphi) - z_2(x)\alpha_1(\varphi)) + z_1(x)(z_2(x)\alpha_o(\varphi) -$$
$$- z_o(x)\alpha_2(\varphi)) + z_2(x)(z_o(x)\alpha_1(\varphi) - z_1(x)\alpha_o(\varphi)) = 0 \ .$$

(ii) a is injective (fibre-wise): If we assumed that this is not the case then there would be some point $x \in \mathbb{P}_2$ and some element $0 \neq \varphi \in \mathbb{C}^n$ such that

$$z_1(x)\alpha_2(\varphi) = z_2(x)\alpha_1(\varphi)$$
$$z_2(x)\alpha_o(\varphi) = z_o(x)\alpha_2(\varphi)$$
$$z_o(x)\alpha_1(\varphi) = z_1(x)\alpha_o(\varphi) \ .$$

But the $\alpha_o(\varphi)$, $\alpha_1(\varphi)$, $\alpha_2(\varphi)$ are linearly dependent, a contradiction to the prestability of α.

(iii) c is surjective: This is proved in the same way by looking at c^T.

(iv) Therefore $\mathcal{F}(\alpha) = \operatorname{Coh}(M(\alpha))$ is a vector bundle of rank $\operatorname{Dim} U - 2n = \operatorname{rank} A(\alpha) - 2n$.

(v) The display of the monad $M(\alpha)$ reads as follows :

$$\begin{array}{ccccccccc} & & & & 0 & & 0 & & \\ & & & & \downarrow & & \downarrow & & \\ 0 & \longrightarrow & \mathbb{C}^n\otimes\mathcal{O}(-1) & \longrightarrow & \mathcal{K} & \longrightarrow & \mathcal{F} & \longrightarrow & 0 \\ & & & & \downarrow & & \downarrow & & \\ 0 & \longrightarrow & \mathbb{C}^n\otimes\mathcal{O}(-1) & \longrightarrow & U\otimes\mathcal{O} & \longrightarrow & \mathcal{P} & \longrightarrow & 0 \\ & & & & \downarrow & & \downarrow & & \\ & & & & \mathbb{C}^n\otimes\mathcal{O}(1) & = & \mathbb{C}^n\otimes\mathcal{O}(1) & & \\ & & & & \downarrow & & \downarrow & & \\ & & & & 0 & & 0 & & \end{array}$$

One immediately computes $c_1(\mathcal{F}) = 0$, $c_2(\mathcal{F}) = n$.

(vi) Since the map $U \to \mathbb{C}^n\otimes V^*$ is injective this implies $h^0(\mathcal{F}) = 0$.

(vii) $h^0(\mathcal{F}^*) = 0$ can be proved by looking at the dual of the display.

1.5. Before proving the main theorem of this section we shall need some more preparations. We shall consider the product $\mathbb{P}_2 \times \mathbb{P}_2$ with projections p and q :

$$\begin{array}{ccc} & \mathbb{P}_2 \times \mathbb{P}_2 & \\ {\scriptstyle p}\swarrow & & \searrow{\scriptstyle q} \\ \mathbb{P}_2 & & \mathbb{P}_2 \end{array} .$$

The diagonal in $\mathbb{P}_2 \times \mathbb{P}_2$ will be denoted by Δ . If $\mathcal{G}$ and $\mathcal{H}$ are two $\mathcal{O}_{\mathbb{P}_2}$-sheaves we define

$$\mathcal{G} \boxtimes \mathcal{H} := p^*\mathcal{G} \otimes_{\mathcal{O}_{\mathbb{P}_2 \times \mathbb{P}_2}} q^*\mathcal{H} .$$

1.5.1. Next we consider the following commutative diagram

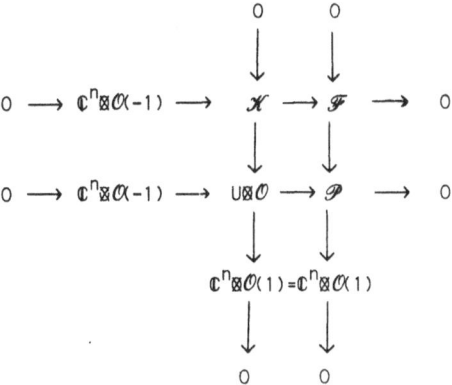

(3)

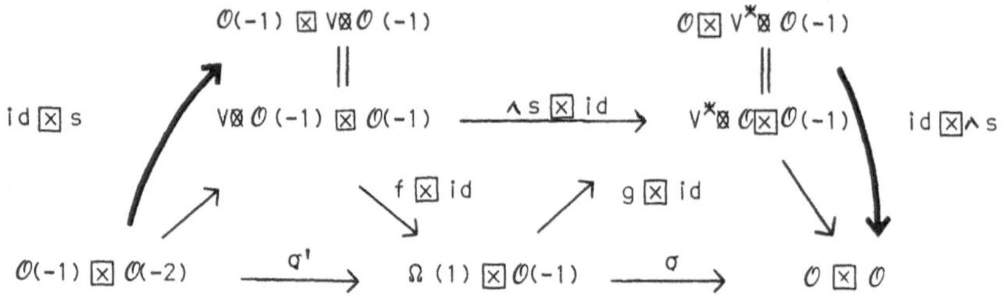

Then σ^T is a section of the bundle $T(-1) \boxtimes \mathcal{O}(1)$. It vanishes transversally on the diagonal Δ . (Under the canonical identifications $H^0(T(-1) \boxtimes \mathcal{O}(1)) = H^0(T(-1)) \boxtimes H^0(\mathcal{O}(1)) = V \boxtimes V^* = \text{Hom}(V,V)$, it can be identified with id_V, and σ' can be identified with $\wedge\sigma$). We get a resolution of the diagonal Δ (see also [10]) by :

(4) $0 \longrightarrow \mathcal{O}(-1) \boxtimes \mathcal{O}(-2) \xrightarrow{\sigma'} \Omega(1) \boxtimes \mathcal{O}(-1) \xrightarrow{\sigma} \mathcal{O} \boxtimes \mathcal{O} \longrightarrow \mathcal{O}_\Delta \longrightarrow 0.$

We can now state the main result of this chapter.

1.5.2. **Theorem.** *By assigning to each pre-stable Kronecker-module α the bundle $\mathcal{F}(\alpha)$ we get a bijection between the equivalence classes of prestable Kronecker-modules $[\alpha]$ with* rank $A(\alpha)$ = 2n+r *and the isomorphism classes of s-stable vector bundles $\mathcal{F}$ of rank r with $c_2(\mathcal{F})$ = n.*

Proof. The map $\alpha \longrightarrow \mathcal{F}(\alpha)$ has been described above. It remains to prove that it is bijective.

(i) the map is surjective : To see this consider any s-stable vector bundle $\mathcal{F}$. We use (4) to get a diagram

$$0 \longrightarrow \mathcal{O}(-1) \boxtimes \mathcal{O}(-2) \xrightarrow{\wedge\sigma} \Omega(1) \boxtimes \mathcal{O}(-1) \xrightarrow{\sigma} \mathcal{O} \boxtimes \mathcal{O} \longrightarrow \mathcal{O}_\Delta \longrightarrow 0$$

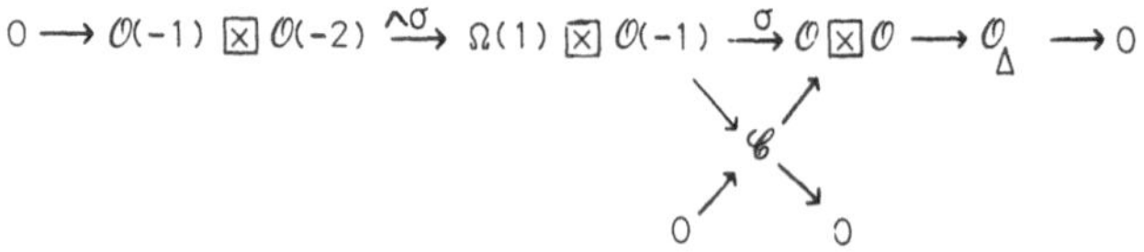

which we can tensor with $p^*\mathcal{F}(-1)$ and get :

$$
\begin{array}{ccccccccc}
0 \longrightarrow \mathcal{F}(-2) \boxtimes \mathcal{O}(-2) & \longrightarrow & \mathcal{F}\otimes\Omega \boxtimes \mathcal{O}(-1) & & \longrightarrow & & \mathcal{F}(-1) \boxtimes \mathcal{O} & \longrightarrow & p^*\mathcal{F}(-1)|\Delta \longrightarrow 0 \\
 & & & \searrow & & \nearrow & & & \\
 & & & & p^*\mathcal{F}\otimes\mathcal{C} & & & & \\
 & & & \nearrow & & \searrow & & & \\
 & & & 0 & & 0 & & &
\end{array}
$$

Applying q_* to the three term exact sequence on the right hand and using $q_*p^*\mathcal{F}(-1)|\Delta = \mathcal{F}(-1)$ we find an exact sequence

$$0 \longrightarrow q_*(p^*(\mathcal{F}(-1)\otimes\mathcal{C}) \longrightarrow q_*p^*\mathcal{F}(-1) \longrightarrow \mathcal{F}(-1) \longrightarrow q_{*1}(p^*\mathcal{F}(-1)\otimes\mathcal{C}) \longrightarrow$$

$$\longrightarrow q_{*1}p^*\mathcal{F}(-1) \longrightarrow 0 .$$

Because of $h^0(\mathcal{F}(-1)) = 0$ it follows that $q_*(p^*\mathcal{F}(-1)\otimes\mathcal{C}) = q_*p^*\mathcal{F}(-1) = 0$ and the above sequence reads

$$(5) \quad 0 \longrightarrow \mathcal{F}(-1) \longrightarrow q_{*1}(p^*\mathcal{F}(-1)\otimes\mathcal{C}) \longrightarrow H^1(\mathcal{F}(-1))\otimes\mathcal{O} \longrightarrow 0 .$$

Now applying q_* to the other exact three term sequence yields

$$0 \longrightarrow q_{*1}(\mathcal{F}(-2)\otimes \mathcal{O}(-2)) \longrightarrow q_{*1}(\mathcal{F}\otimes\Omega \boxtimes \mathcal{O}(-1)) \longrightarrow q_{*1}(p^*\mathcal{F}(-1)\otimes\mathcal{C}) \longrightarrow$$

$$\longrightarrow q_{*2}(\mathcal{F}(-2) \boxtimes \mathcal{O}(-2)).$$

The last term vanishes because of $h^2(\mathcal{F}(-2)) = h^0(\mathcal{F}^*(-1)) = 0$, and one has

$$(6) \quad 0 \longrightarrow H^1(\mathcal{F}(-2))\otimes\mathcal{O}(-2) \longrightarrow H^1(\mathcal{F}\otimes\Omega)\otimes\mathcal{O}(-1) \longrightarrow q_{*1}(p^*\mathcal{F}(-1)\otimes\mathcal{C}) \longrightarrow 0 .$$

Tensoring (5) and (6) with $\mathcal{O}(1)$ we can combine them to get the following commutative diagramm

$$
\begin{array}{ccccccc}
 & & 0 & & 0 & & \\
 & & \downarrow & & \downarrow & & \\
0 \longrightarrow H^1(\mathcal{F}(-2))\otimes\mathcal{O}(-1) & \longrightarrow & \mathcal{K} & \longrightarrow & \mathcal{F} & \longrightarrow & 0 \\
 & & \downarrow & & \downarrow & & \\
0 \longrightarrow H^1(\mathcal{F}(-2))\otimes\mathcal{O}(-1) & \longrightarrow & H^1(\mathcal{F}\otimes\Omega)\otimes\mathcal{O} & \longrightarrow & q_{*1}(p^*\mathcal{F}\otimes\mathcal{C})\otimes\mathcal{O}(1) & \longrightarrow & 0 \\
 & & \downarrow & & \downarrow & & \\
 & & H^1(\mathcal{F}(-1))\otimes\mathcal{O}(1) & = & H^1(\mathcal{F}(-1))\otimes\mathcal{O}(1) & & \\
 & & \downarrow & & \downarrow & & \\
 & & 0 & & 0 & &
\end{array}
$$

But this means that $\mathcal{F}$ belongs to the monad

$$H^1(\mathcal{F}(-2))\otimes\mathcal{O}(-1) \longrightarrow H^1(\mathcal{F}\otimes\Omega)\otimes\mathcal{O} \longrightarrow H^1(\mathcal{F}(-1))\otimes\mathcal{O}(1)$$

which in view of (3) can be written as

$$\begin{array}{ccccccc} & & H^1(\mathcal{F}-2))\otimes V\otimes\mathcal{O} & \xrightarrow{A(\alpha_{\mathcal{F}})} & H^1(\mathcal{F}-1))\otimes V^*\otimes\mathcal{O} & & \\ & {}^{id\otimes s}\nearrow & & \searrow^{F} \quad {}^{G}\nearrow & & \searrow^{id\otimes\wedge s} & \\ H^1(\mathcal{F}-2))\otimes\mathcal{O}(-1) & \longrightarrow & & H^1(\mathcal{F}\otimes\Omega)\otimes\mathcal{O} & & \longrightarrow & H^1(\mathcal{F}-1))\otimes\mathcal{O}(1) \end{array} .$$

In other words : $\mathcal{F}$ is the cohomology bundle of the monad $M(\alpha_{\mathcal{F}})$.

(ii) this map is injective. Let us assume that α and α' give isomorphic bundles $\mathcal{F}$ and $\mathcal{F}'$. Then the isomorphic bundles $\mathcal{F}\oplus(n-r)\mathcal{O}$ and $\mathcal{F}'\oplus(n-r)\mathcal{O}$ belong to the monads

$$\mathbb{C}^n\otimes\mathcal{O}(-1) \longrightarrow \mathbb{C}^n\otimes V\otimes\mathcal{O} \xrightarrow{A(\alpha)} \mathbb{C}^n\otimes V^*\otimes\mathcal{O} \longrightarrow \mathbb{C}^n\otimes\mathcal{O}(1)$$

and

$$\mathbb{C}^n\otimes\mathcal{O}(-1) \longrightarrow \mathbb{C}^n\otimes V\otimes\mathcal{O} \xrightarrow{A(\alpha')} \mathbb{C}^n\otimes V^*\otimes\mathcal{O} \longrightarrow \mathbb{C}^n\otimes\mathcal{O}(1) .$$

By [4, Prop.4] it follows that there is an isomorphism of monads :

$$\begin{array}{ccccc} \mathbb{C}^n\otimes\mathcal{O}(-1) & \longrightarrow & \mathbb{C}^n\otimes V^*\otimes\mathcal{O} & \longrightarrow & \mathbb{C}^n\otimes\mathcal{O}(1) \\ \downarrow{\scriptstyle \omega_2\otimes id} & & \downarrow{\scriptstyle \sigma\otimes id} & & \uparrow{\scriptstyle \omega_1^T\otimes id} \\ \mathbb{C}^n\otimes\mathcal{O}(-1) & \longrightarrow & \mathbb{C}^n\otimes V^*\otimes\mathcal{O} & \longrightarrow & \mathbb{C}^n\otimes\mathcal{O}(1) \end{array}$$

But then $\sigma = \omega_1^T \otimes id_{V^*}$ and we have a commutative diagram

$$\begin{array}{ccc} \mathbb{C}^n\otimes V & \xrightarrow{A(\alpha)} & \mathbb{C}^n\otimes V^* \\ \downarrow{\scriptstyle \omega_2\otimes id} & & \uparrow{\scriptstyle \omega_1^T\otimes id} \\ \mathbb{C}^n\otimes V & \xrightarrow{A(\alpha')} & \mathbb{C}^n\otimes V^* \end{array}$$

from which we conclude $\omega_1^T\ \alpha_i'\ \omega_2 = \alpha_i$ for $i=0,1,2$; i.e. $[\alpha] = [\alpha']$.

1.5.3. Remark. The proof given here goes back to the spectral sequence argument given by Beilinson [5] and also used by Le Potier [10] .

1.6. Let $W: = \mathrm{Hom}(\mathbb{C}^n\otimes V^*,\mathbb{C}^n)$ be the vector space of Kronecker-modules.

We define

$\tilde{M}(r;o,n) := \{\alpha \in W \; ; \; \alpha \text{ is pre-stable with rank } A(\alpha) = 2n+r\}$.

Then $\tilde{M}(r,0,n)$ is a quasi-affine variety since the rank condition is Zariski-closed whereas pre-stability is an open condition.

1.6.1. Proposition. *On $\mathbb{P}_2 \times \tilde{M}(r,0,n)$ exists a "universal bundle", i.e. a vector bundle $\mathcal{E}$, such that $\mathcal{E}|\mathbb{P}_2 \times \{\alpha\} = \mathcal{F}(\alpha)$. In particular this means that the family of s-stable rank-r bundles with $c_1(\mathcal{F}) = 0$ and $c_2(\mathcal{F}) = n$ is bounded.*

Proof. We consider the product $\mathbb{P}_2 \times \tilde{M}(r;0,n)$ with projections p_1 and p_2 :

$$\begin{array}{ccc} & \mathbb{P}_2 \times \tilde{M}(r,0,n) & \\ {}^{p_1}\swarrow & & \searrow^{p_2} \\ \mathbb{P}_2 & & \tilde{M}(r;0,n) \end{array}$$

On $\tilde{M}(r,0,n)$ we define a bundle map

$$A' : \mathbb{C}^n \otimes V \otimes \mathcal{O}_{\tilde{M}} \longrightarrow \mathbb{C}^n \otimes V^* \otimes \mathcal{O}_{\tilde{M}}$$

by $$A'|\{\alpha\} := A(\alpha) .$$

Next we consider the morphism

$$\tilde{A} := p_2^*(A') : \mathbb{C}^n \otimes V \otimes \mathcal{O}_{\tilde{M} \times \mathbb{P}_2} \longrightarrow \mathbb{C}^n \otimes V^* \otimes \mathcal{O}_{\tilde{M} \times \mathbb{P}_2}$$

and we define maps

$$\tilde{a} : \mathbb{C}^n \otimes p_1^* \mathcal{O}(-1) \xrightarrow{p_1^*(s)} \mathbb{C}^n \otimes V \otimes \mathcal{O}_{\tilde{M} \times \mathbb{P}_2} \xrightarrow{\tilde{A}} \mathbb{C}^n \otimes V^* \otimes \mathcal{O}_{\tilde{M} \times \mathbb{P}_2}$$

$$\tilde{c} : \mathbb{C}^n \otimes V^* \otimes \mathcal{O}_{\tilde{M} \times \mathbb{P}_2} \xrightarrow{p_1^*(\wedge s)} \mathbb{C}^n \otimes p_1^* \mathcal{O}(1) .$$

Since $\tilde{A}$ has constant rank on all fibres the image $\mathcal{G} := \operatorname{Im}(\tilde{A})$ is a vector bundle again. We get a "universal monad"

$$\tilde{M} : \mathbb{C}^n \otimes p_1^* \mathcal{O}(-1) \xrightarrow{\tilde{a}} \mathcal{G} \xrightarrow{\tilde{c}} \mathbb{C}^n \otimes p_1^* \mathcal{O}(1)$$

whose cohomology $\mathcal{E}$ is the desired universal bundle.

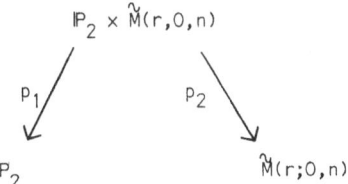

1.6.2. Remarks. (i) Maruyama [7, Theorem 2.8] has proved that stability is an open property. Hence there is a Zariski-open set

$$\tilde{M}_s(r,0,n) \subseteq \tilde{M}(r,0,n)$$

consisting of all those α for which $\mathcal{F}(\alpha)$ is stable.

(ii) Let $M(r;0,n)$ be the moduli-scheme of stable rank-r vector bundles with $c_1(\mathcal{F}) = 0$ and $c_2(\mathcal{F}) = n$. By its universal property we have an epimorphism

$$\pi : \tilde{M}_s(r;0,n) \longrightarrow M(r;0,n), \qquad \alpha \rightarrow \mathcal{F}(\alpha).$$

1.7. We want to conclude this chapter with some remarks concerning the geometric meaning which $\alpha_{\mathcal{F}}$ has for the vector bundle $\mathcal{F}$.

1.7.1. We define

$$\Delta(\alpha) := \Delta(\alpha_{\mathcal{F}}) := \det(\alpha_{\mathcal{F}}(z)) \in \Gamma(\mathcal{O}_{\mathbb{P}_2^*}(n))$$

and

$$S(\mathcal{F}) := \{\Delta(\alpha) = 0\} = \{z;\ \det(\alpha(z)) = 0\} \quad .$$

The discriminant $\Delta(\alpha)$ is a homogeneous polynomial of degree n (up to a scalar it is uniquely determined by $[\alpha]$), and hence $S(\mathcal{F}) \subseteq \mathbb{P}_2^*$ will be a curve of degree n or the whole plane. The meaning of $S(\mathcal{F})$ is

$$S(\mathcal{F}) = \{ L \in \mathbb{P}_2^* ;\ \mathcal{F}|L \neq r.\, \mathcal{O}_L \} .$$

To see this consider a line L with equation $\{z=0\}$. The exact sequence

$$0 \longrightarrow \mathcal{F}(-2) \xrightarrow{z} \mathcal{F}(-1) \longrightarrow \mathcal{F}(-1)|L \longrightarrow 0$$

yields

$$0 \longrightarrow H^0(\mathcal{F}-1)|L) \longrightarrow H^1(\mathcal{F}-2)) \xrightarrow{\alpha(z)} H^1(\mathcal{F}-1)).$$

Hence $\det \alpha(z) = 0$ if and only if $h^0(\mathcal{F}-1)|L \neq 0$, but this is equivalent to $\mathcal{F}|L \neq r.\, \mathcal{O}_L$.

In the rank-2 case $S(\mathcal{F})$ is nothing but the curve of jumping lines (cf. [2]), whereas for higher rank the determinant $\Delta(\alpha)$ may vanish identically (see example (1.7.4) below).

1.7.3. For any $\alpha = \alpha_{\mathcal{F}}$ with $\Delta(\alpha) \not\equiv 0$ we define a map A as follows :

$$A: \mathbb{C}^n \otimes \mathcal{O}_{\mathbb{P}_2^*}(-1) \xrightarrow{s} \mathbb{C}^n \otimes V^* \otimes \mathcal{O}_{\mathbb{P}_2^*} \xrightarrow{\alpha} \mathbb{C}^n \otimes \mathcal{O}_{\mathbb{P}_2^*} .$$

Over a point $x \in \mathbb{P}_2^*$ with coordinate vector z the map A is just $\alpha(z)$.

We define a sheaf $\mathcal{L}$ by

$$0 \longrightarrow \mathbb{C}^n \otimes \mathcal{O}(-1) \xrightarrow{A} \mathbb{C}^n \otimes \mathcal{O} \longrightarrow \mathcal{L} \longrightarrow 0 .$$

The sheaf $\mathcal{L}$ has $S(\mathcal{F})$ as its support and has moreover the following properties :

(i) $\mathcal{L}$ is an $\mathcal{O}_{S(\mathcal{F})}$-sheaf

(ii) $h^0(\mathcal{L}(-1)) = h^1(\mathcal{L}(-1)) = 0$

(iii) If $\mathcal{L}$ is invertible on $S(\mathcal{F})$ then $\deg(\mathcal{L}(-1)) = \frac{1}{2} \deg \omega_{S(\mathcal{F})}$.

(iv) The pair $(S(\mathcal{F}), \mathcal{L})$ determines $\mathcal{F}$ uniquely.

On the other hand if $\deg(S(\mathcal{F})) = n$ and if $\mathcal{L} \in \mathrm{Pic}(S(\mathcal{F}))$ is of degree $\frac{1}{2} n(n-3)$ with $h^0(\mathcal{L}(-1)) = h^1(\mathcal{L}(-1)) = 0$, then one can reconstruct a Kronecker-module α (but perhaps not a bundle) from it.

The sheaf $\mathcal{L}$ was used in [2] to classify rank-2 bundles. There, since such bundles carry a canonical symplectic form, the sheaf $\mathcal{L}(-1)$, if it was invertible and if $S(\mathcal{F})$ was reduced, turned out to be a θ-characteristic.

1.7.3. If $\mathcal{F}$ carries a symplectic (orthogonal) form then one can find a representative α of $[\alpha_{\mathcal{F}}]$ such that the matrices α_i are symmetric (skew-symmetric). In the rank-2 case this leads to the nets of quadrics considered by Barth in [2] .

1.7.4. If $c_2 = 3$ we have two possibilities :

$$\operatorname{rank} A(\alpha) = \begin{cases} 9 \\ 8 \end{cases} \qquad \text{and } \operatorname{rank}(\mathcal{F}) = \begin{cases} 3 \\ 2 \end{cases} .$$

Here $\Delta(\alpha)$ will be a cubic polynomial and there is exactly one possibility that it vanishes identically :

Proposition. *If $\mathcal{F}$ is s-stable with $c_2(\mathcal{F}) = 3$ and if $\Delta(\alpha) \equiv 0$ then $\mathcal{F} = S^2T(-3)$.*

Proof. Let $\mathcal{F}$ be any such bundle and let α be a representative of $\alpha_{\mathcal{F}}$. By our assumption we have vectors $0 \neq \varphi_i \in \operatorname{Ker} \alpha_i$. Because of pre-stability any two of the vectors φ_i will be linearly independent.

Case 1. $\varphi_0, \varphi_1, \varphi_2$ form a basis of $\mathbb{C}^3$. We can assume (by prestability) that one of the matrices α_i has rank two. We shall consider the case that this is fulfilled by α_0 (the other cases can be worked in the same way). Then we may choose coordinates such that

$$\alpha_0 = \begin{pmatrix} 0 & 0 & 0 \\ 0 & 0 & 1 \\ 0 & -1 & 0 \end{pmatrix} \quad \alpha_1 = \begin{pmatrix} a_1 & 0 & b_1 \\ a_2 & 0 & b_2 \\ a_3 & 0 & b_3 \end{pmatrix} \quad \alpha_2 = \begin{pmatrix} c_1 & d_1 & 0 \\ c_2 & d_2 & 0 \\ c_3 & d_3 & 0 \end{pmatrix}.$$

It follows immediately from $\Delta(\alpha) \equiv 0$ that $a_1 = c_1 = 0$. But prestability then implies $b_1 \neq 0 \neq d_1$. Moreover we can change our coordinates such that

$$\alpha_0 = \begin{pmatrix} 0 & 0 & 0 \\ 0 & 0 & 1 \\ 0 & -1 & 0 \end{pmatrix} \quad \alpha_1 = \begin{pmatrix} 0 & 0 & b_1' \\ a_2' & 0 & b_2' \\ a_3' & 0 & b_3' \end{pmatrix} \quad \alpha_2 = \begin{pmatrix} 0 & 1 & 0 \\ c_2' & 0 & 0 \\ c_3' & 0 & 0 \end{pmatrix}.$$

Again exploiting $\Delta(\alpha) \equiv 0$ we find $a_2' = b_2' = b_3' = c_3' = 0$ and $a_3' - c_2' b_1' = 0$. Next we can normalize $c_2' = -1$, and one easily sees that the then found normal form is equivalent to

$$\alpha_0 = \begin{pmatrix} 0 & 0 & 0 \\ 0 & 0 & 1 \\ 0 & -1 & 0 \end{pmatrix} \quad \alpha_1 = \begin{pmatrix} 0 & 0 & 1 \\ 0 & 0 & 0 \\ -1 & 0 & 0 \end{pmatrix} \quad \alpha_2 = \begin{pmatrix} 0 & 1 & 0 \\ -1 & 0 & 0 \\ 0 & 0 & 0 \end{pmatrix}$$

Case 2. $\varphi_0, \varphi_1, \varphi_2$ are linearly dependent. Then we can assume

$$\varphi_0 = \begin{pmatrix} 1 \\ 0 \\ 0 \end{pmatrix} \qquad \varphi_1 = \begin{pmatrix} 0 \\ 1 \\ 0 \end{pmatrix} \qquad \varphi_2 = \begin{pmatrix} 1 \\ 1 \\ 0 \end{pmatrix}.$$

By similar arguments as above and by exploiting $\Delta(\alpha) \equiv 0$ one finds coordinates such that

$$\alpha_0 = \begin{pmatrix} 0 & 0 & 0 \\ 0 & 0 & 0 \\ 0 & 1 & a_3 \end{pmatrix} \quad \alpha_1 = \begin{pmatrix} 0 & 0 & 0 \\ 1 & 0 & b_2 \\ 0 & 0 & 0 \end{pmatrix} \quad \alpha_2 = \begin{pmatrix} 1 & -1 & c_1 \\ 0 & 0 & 0 \\ 0 & 0 & 0 \end{pmatrix}$$

But then $\alpha_0^T \begin{pmatrix} 1 \\ 0 \\ 0 \end{pmatrix} = \alpha_1^T \begin{pmatrix} 1 \\ 0 \\ 0 \end{pmatrix} = \begin{pmatrix} 0 \\ 0 \\ 0 \end{pmatrix}$ which contradicts pre-stability.

It follows that the bundle $\mathscr{F}$ is unique, and hence must be $S^2T(-3)$.

Remarks. $S^2T(-3)$ is homogeneous with $\mathscr{F}|L = \mathcal{O}_L(-1) \oplus \mathcal{O}_L \oplus \mathcal{O}_L(1)$ for every line L. In addition it carries an orthogonal structure.

1.7.5. If $c_2 = 4$ the rank of $A(\alpha_{\mathscr{F}})$ can be 12, 11 or 10. The rank of the corresponding bundle will then be 4, 3 or 2. Whereas for $r = 4$ the curve $S(\mathscr{F})$ can be a general quartic, one finds for $r = 2$ that $S(\mathscr{F})$ must be a Lüroth quartic (see [2]).

2. IRREDUCIBILITY OF $M(r;0,n)$

2.1. By $M(r;0,n)$ we denote the moduli-scheme of stable rank-r bundles $\mathscr{F}$ with $c_1(\mathscr{F}) = 0$ and $c_2(\mathscr{F}) = n$. It is well known that stable vector bundles are simple, i.e. that $h^0(\mathscr{F} \otimes \mathscr{F}^*) = 1$. By Serre-duality this implies $h^2(\mathscr{F} \otimes \mathscr{F}^*) = 0$ and one can use Riemann-Roch to find

$$h^1(\mathscr{F} \otimes \mathscr{F}^*) = 1 - \chi(\mathscr{F} \otimes \mathscr{F}^*) = 2rn - r^2 + 1.$$

From the above follows that the moduli-scheme will (unless it is empty) be smooth and all components will have dimension $2rn-r^2+1$.

The main point of this chapter is to prove

2.2. Theorem. *The varieties* $M(r;0,n)$ *are irreducible.*

2.3. Before we can give the proof it will be necessary to collect some lemmas.

2.3.1. Definition. A matrix $a \in M(n\times n,\mathbb{C})$ will be called *regular* if its minimal polynomial has degree n.

Remarks. (i) Our notion of regular means that a is a regular element in the Lie algebra $\mathfrak{gl}(n,\mathbb{C})$, i.e. that its centralizer has dimension n. We hope that no confusion arises with the notion of a regular matrix which means $\det(a) \neq 0$.

(ii) Another equivalent condition is the following : If the Jordan form of a is given by

$$\begin{bmatrix} \lambda_1 & & & & & \\ & \ddots & & & & \\ & & \lambda_t & & & \\ & & & \boxed{\begin{matrix}\lambda_{t+1} & 1 & & \\ & \ddots & \ddots & \\ & & \ddots & 1 \\ & & & \lambda_{t+1}\end{matrix}} & & \\ & & & & \ddots & \\ & & & & & \boxed{\begin{matrix}\lambda_{m} & 1 & & \\ & \ddots & \ddots & \\ & & \ddots & 1 \\ & & & \lambda_{m}\end{matrix}} \end{bmatrix}$$

then all λ_i are different.

2.3.2. Definition. For any matrix $b \in M(n\times n,\mathbb{C})$ we define a variety

$$C_b := \{c; c = [a,b] \text{ for some } a, \operatorname{rank}(c) = r\} .$$

2.3.3. Lemma. *If* b *is regular then*

$$C_b = \{c; \operatorname{tr}(c) = \operatorname{tr}(bc) = \ldots = \operatorname{tr}(b^{n-1}c) = 0, \operatorname{rank}(c) = r\}$$

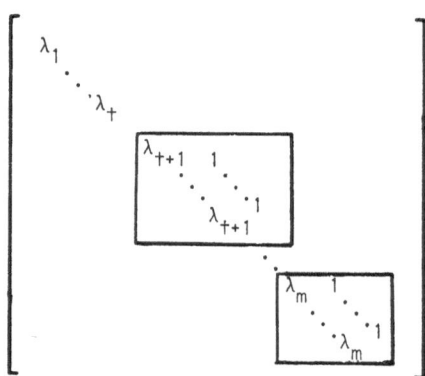

Proof. It will be sufficient to prove

$$C_b' := \{c; c = [a,b]\} = \{c; tr(c) = \ldots = tr(b^{n-1}c) = 0\} .$$

C_b' is the image of the vector space homomorphism

$$M(n\times n,\mathbb{C}) \to M(n\times n,\mathbb{C}), \quad a \mapsto [a,b] .$$

Since b is regular the defect of this map is n, and C_b' is a linear subspace of codimension n. For any ν we have

$$tr(b^\nu[a,b]) = tr[a,b^{\nu+1}] = 0 .$$

On the other hand, since b is regular, the matrices $\mathbb{1},\ldots b^{n-1}$ are linearly independent. Hence the equations $tr(c) = \ldots = tr(b^{n-1}c) = 0$ describe a subspace of codimension n which then must be C_b'.

Remark. If $b_o,\ldots,b_{n-1}$ is any basis of the centralizer of b then by the above clearly

$$C_b = \{c; tr(b_o c) = \ldots = tr(b_{n-1}c) = 0, \ rank(c) = r\}.$$

The most important technical lemma for our proof will be

2.3.4. Lemma. *If* b *is regular then* C_b *is irreducible and has dimension* $2rn-r^2-n$.

Proof. (i) We set $\mathbb{1}_r := \begin{pmatrix} E_r & 0 \\ 0 & 0 \end{pmatrix}$ where E_r is the $(r\times r)$-unit matrix, and consider the map

$$\Phi : GL(n,\mathbb{C}) \times GL(n,\mathbb{C}) \longrightarrow M(n\times n,\mathbb{C})$$
$$(g,h) \longrightarrow g\,\mathbb{1}_r h .$$

Its image will be the $(2rn-r^2)$-dimensional variety of matrices having rank r. All fibres of this map have dimension $n^2+(n-r)^2$. It will now be sufficient to prove that

$$\mathcal{C}_b := \Phi^{-1}(C_b)$$

is irreducible and of dimension $2n^2-n$.

(iii) We first assume that b is diagonalizable ; i.e. we may assume that

$$b = \begin{pmatrix} \lambda_1 & & \\ & \ddots & \\ & & \lambda_n \end{pmatrix}$$

with all λ_i different. Then a basis of the centralizer is given by

$$b_0 = \begin{pmatrix} 1 & & & \\ & 0 & & \\ & & \ddots & \\ & & & 0 \end{pmatrix}, \ldots, \quad b_{n-1} = \begin{pmatrix} 0 & & & \\ & 0 & & \\ & & \ddots & \\ & & & 1 \end{pmatrix}$$

For any matrix c we have

$$tr(b_i c) = c_{ii} \; .$$

$\tilde{\mathcal{C}}_b$ is therefore given by the equations

$$c_{ii} = \sum_{k=1}^{r} g_{ik} h_{ki} = 0 \; .$$

By $V_i \subseteq M(nxn,\mathbb{C}) \times M(nxn,\mathbb{C})$ we denote the subspace with coordinates g_{ij}, h_{ji}; $1 \leq j \leq n$. Then $M(nxn,\mathbb{C}) \times M(nxn,\mathbb{C}) = \bigoplus_{i=1}^{n} V_i$ and each equation $c_{ii} = 0$ defines in V_i a quadric Q_i of rank 2r. Since $r \geq 2$ these quadrics are irreducible and the same must hold for

$$\tilde{\mathcal{C}}_b = (Q_1 x \ldots x \; Q_n) \cap GL(n,\mathbb{C}) \times GL(n,\mathbb{C})) \; ,$$

which is a variety of dimension $2n^2-n$.

(iii) To tackle the general case we start with the following remark : Consider some (sxs)-matrix

$$b' = \begin{pmatrix} \lambda & 1 & & \\ & \ddots & \ddots & \\ & & \ddots & 1 \\ & & & \lambda \end{pmatrix}$$

A basis of the centralizer of b' is then given by

$$b'_0 = \begin{pmatrix} 1 & & \\ & \ddots & \\ & & 1 \end{pmatrix}, \quad b'_1 = \begin{pmatrix} 0 & 1 & & \\ & \ddots & \ddots & \\ & & \ddots & 1 \\ & & & 0 \end{pmatrix} \quad , \ldots, \quad b'_{s-1} = \begin{pmatrix} 0 & & 1 \\ & \ddots & \\ & & 0 \end{pmatrix}$$

For any (sxs)-matrix c' we have

$$tr(b'_\nu c') = \sum_{i=1}^{s-\nu} c'_{i+\nu,i}$$

Now assume that b is a Jordan matrix containing such a block b', i.e.

$$b = \begin{bmatrix} \boxed{b'} & & \\ & \ddots & \\ & & \boxed{\ddots} \end{bmatrix}$$

We define

$$b_i := \begin{bmatrix} \boxed{b_i'} & & \\ & \ddots o & \\ & & \ddots o \end{bmatrix}$$

For $c = g \, 1\!\!1_r h$ the equations $tr(b_\nu . c)$; $\nu = o, \dots, s-1$ are given by

$$f_\nu := \sum_{i=1}^{s-\nu} \sum_{k=1}^{r} g_{i+\nu,k} h_{ki} = o.$$

We take $V_1^{(s)} \subseteq M(n\times n,\mathbb{C}) \times M(n\times n,\mathbb{C})$ to be the subspace given by the coordinates g_{ij}; $1 \leq i \leq s$, $1 \leq j \leq n$ and h_{ij}; $1 \leq i \leq n$, $1 \leq j \leq s$. We assert that the variety $R := R_1^{(s)} \subseteq V_1^{(s)}$ given by the equations f_ν is irreducible and of codimension s.

To see this consider the functional matrix

$$J := \left(\frac{\partial f_\nu}{\partial g_{\ell m}} , \frac{\partial f_\nu}{\partial h_{pq}} \right) .$$

One easily cheks that

$$\frac{\partial f_\nu}{\partial g_{\ell m}} = \begin{cases} h_{m,\ell-\nu} & \text{if } 1 \leq m \leq r, \quad \ell \geq \nu+1 \\ 0 & \text{otherwise} \end{cases}$$

$$\frac{\partial f_\nu}{\partial h_{pq}} = \begin{cases} g_{q+\nu,p} & \text{if } q \leq n-\nu, \ 1 \leq p \leq r \\ 0 & \text{otherwise .} \end{cases}$$

Hence, for fixed m,p we get matrices

$$\left(\frac{\partial f_\nu}{\partial g_{\ell m}}\right)_{\nu,\ell} = \begin{pmatrix} h_{m1} & \cdots & h_{ms} \\ & \ddots & \vdots \\ & & h_{m1} \end{pmatrix}$$

$$\left(\frac{\partial f_\nu}{\partial h_{pq}}\right)_{\nu,q} = \begin{pmatrix} g_{1p} & \cdots & g_{sp} \\ \vdots & ⋰ & \\ g_{sp} & & \end{pmatrix}$$

From this one easily deduces that R is of codimension s everywhere and that its singular locus is given by

$$S(R) = \{f_\nu = h_{m1} = g_{sp} = 0 \; ; \; 0 \leq \nu \leq s; \; 1 \leq m, \, p \leq r\} \, .$$

Now looking at the functional matrix of the above equations we find

$$\operatorname{codim}_R S(R) \geq 2r-2.$$

In order to prove that R is irreducible we have to prove that R-S(R) is connected. But this follows, since $r \geq 2$, from the above by the well theorem of Hartshorne that complete intersections are connected in codimension two.

(iv) We finally can prove the lemma as follows : We assume

$$b = \begin{bmatrix} \lambda_1 & & & & & \\ & \ddots & & & & \\ & & \lambda_t & & & \\ & & & \boxed{\begin{matrix} \lambda_{t+1} & 1 & \\ & \ddots & \ddots & 1 \\ & & & \lambda_{t+1} \end{matrix}} & & \\ & & & & \ddots & \\ & & & & & \boxed{\begin{matrix} \lambda_m & 1 & \\ & \ddots & \ddots \\ & & \lambda_m & 1 \end{matrix}} \end{bmatrix}$$

If the size of the Jordan block belonging to λ_μ is $s(\mu)$ then

$$\tilde{C}_b = (Q_1 \times \ldots \times Q_t \times R_{t+1}^{s(t+1)} \ldots \times R_m^{s(m)}) \cap (GL(n,\mathbb{C}) \times GL(n,\mathbb{C}))$$

and we are done by (ii) and (iii).

2.3.5. **Lemma.** *The variety*

$$\hat{M}^o := \{(a,b) \; ; \; \text{rank } [a,b] = r \text{ and } b \text{ is regular}\}$$

is irreducible and has dimension $2rn-r^2+n^2$.

Proof. By $M_{reg} \subsetneq M(n\times n,\mathbb{C})$ we denote the Zariski-open set of all regular matrices. We have a map

$$p: \hat{M}^o \longrightarrow M_{reg}, \; (a,b) \longmapsto b.$$

Our assertion will follow from :

(i) p is surjective.

(ii) All fibres are irreducible and have dimension $2rn-r^2$.

(iii) $\hat{M}^o$ is equi-dimensional.

Now (i) follows immediately from $C_b \neq \emptyset$. To prove (ii) we consider a fibre $F_b := \{a \; ; \; \text{rank } [a,b] = r\}$. Then we have an epimorphism

$$q: F_b \longrightarrow C_b \; ; \; a \longmapsto [a,b].$$

Since b is regular, F_b is an affine $\mathbb{C}^n$-bundle over C_b and (ii) follows from (2.3.4). Couting equations one sees that $\dim W \geq 2rn-r^2+n^2$ for every component W of $\hat{M}^o$. But from (i) and (ii) follows $\dim W \leq 2rn-r^2+n^2$, and hence we are done.

2.3.6. **Proposition.** *The variety*

$$\hat{M} := \{(a,b) \; ; \; \text{rank } [a,b] = r\}$$

is irreducible.

Proof. In view of (2.3.5.) it will suffice to prove $\overline{\hat{M}^o} = \hat{M}$. Take a pair (a,b) with rank $[a,b] = r$ where b is not regular. One can always find a regular matrix b' such that $[a,b'] = 0$. Then $b + tb'$ will be regular for all but finitely many t and $[a,b+tb'] = [a,b]$.

2.4. Before finally proving the theorem we still need a lemma.

2.4.1. Definition. $M^o(r;o,n) = \{\mathcal{F}\in M(r;o,n)\ ;\ \mathcal{F}|L = r.\mathcal{O}_L$ for some $L\}$

2.4.1. Lemma (Hirschowitz) : $\overline{M^o(r;o,n)} = M(r;o,n)$.

Proof. Take any stable bundle $\mathcal{F}$ not belonging to $M^o(r;o,n)$. We have to find a small deformations $\mathcal{F}'$ of $\mathcal{F}$ which is trivial on some line L_o. The exact sequence

$$0 \longrightarrow \mathcal{F}\otimes\mathcal{F}^*(-1) \longrightarrow \mathcal{F}\otimes\mathcal{F}^* \longrightarrow \mathcal{F}\otimes\mathcal{F}^*|L_o \longrightarrow 0$$

yields

$$H^1(\mathcal{F}\otimes\mathcal{F}^*) \longrightarrow H^1(\mathcal{F}\otimes\mathcal{F}^*|L_o) \longrightarrow H^2(\mathcal{F}\otimes\mathcal{F}^*(-1)) .$$

On the line L_o we can find a small deformation of $\mathcal{F}|L_o$ which is trivial. Since $\mathcal{F}$ is stable it follows that $h^2(\mathcal{F}\otimes\mathcal{F}^*(-1)) = h^o(\mathcal{F}\otimes\mathcal{F}^*(-2)) = 0$, and therefore a deformation already exists on $\mathbb{P}_2$.

2.5. **Proof of** (2.2) : (i) In (1.6) we introduced the vector space W of all Kronecker-modules, and defined $\tilde{M}(r;o,n)$ as the variety of pre-stable Kronecker-modules with rank $A(\alpha) = 2n+r$. By $\tilde{M}_s(r;o,n)$ we denoted the Zariski-open set of those α for which $\mathcal{F}(\alpha)$ is stable, and we had an epimorphism

$$\pi : \tilde{M}_s(r;o,n) \longrightarrow M(r;o,n) .$$

In view of (2.4.2.) it suffices to prove irreducibility of

$$\tilde{M}^o := \pi^{-1}(M^o(r;o,n)) = \{\alpha\in\tilde{M}_s(r;o,n)\ ;\ \Delta(\alpha) \not\equiv 0\}.$$

But since $\tilde{M}_s(r;o,n)$ was open and since pre-stability too is an open condition it is enough to prove irreducibility of

$$\check{M}^o := \{\alpha\in W\ ;\ \text{rank } A(\alpha) = 2n+r \text{ and } \Delta(\alpha) \not\equiv 0\}.$$

(ii) For all $z \in V^*$ we define

$$\check{M}^o_z := \{\alpha \in \check{M}^o\ ;\ \det(\alpha(z)) \neq 0 .$$

Since $\bigcup_{z\in V^*} \check{M}^o_z = \check{M}^o$, and since $\check{M}^o_z \cap \check{M}^o_{z'} \neq \emptyset$ the problem reduces to prove irreducibility of $\check{M}^o_{z_o}$.

(iii) The map

$$q : \check{M}^o_{z_o} \longrightarrow GL(n,\mathbb{C}), \quad \alpha \longmapsto \alpha_o$$

is surjective. The group $G = GL(n,\mathbb{C}) \times GL(n,\mathbb{C})$ operates on $\check{M}^o_{z_o}$ and on $GL(n,\mathbb{C})$. This operation is transitive on $GL(n,\mathbb{C})$ and q is G-equivariant. Hence it will do to prove the irreducibility of the fibre $q^{-1}(\mathbb{1})$.

(iv) We shall finish the proof by showing $q^{-1}(\mathbb{1}) = \hat{M}$ and applying (2.3.6.). To see this latter equation it remains to prove that

$$\operatorname{rank} \begin{pmatrix} o & b & -a \\ -b & o & \mathbb{1} \\ a & -\mathbb{1} & o \end{pmatrix} = 2n + \operatorname{rank} [a,b] .$$

Now the rank of this matrix equals the rank of

$$\begin{pmatrix} \mathbb{1} & a & b \\ o & \mathbb{1} & o \\ o & o & \mathbb{1} \end{pmatrix} \circ \begin{pmatrix} o & b & -a \\ -b & o & \mathbb{1} \\ a & -\mathbb{1} & o \end{pmatrix} = \begin{pmatrix} -[a,b] & o & o \\ * & o & \mathbb{1} \\ * & -\mathbb{1} & o \end{pmatrix}$$

and from this our assertion follows.

3. On stability.

3.1. We first want to recollect the definition of stability due to Mumford-Takemoto [11].

3.1.1. Definition. A rank-r vector bundle $\mathcal{F}$ is *stable (semi-stable) in the sense of Mumford-Takemoto* if for all coherent non-torsion quotients $\mathcal{G}$ of $\mathcal{F}$ with $0 < s := \operatorname{rank} \mathcal{G} < r$ the following holds :

$$\frac{c_1(\mathcal{G})}{s} > \frac{c_1(\mathcal{F})}{r} \quad \left(\text{or } \frac{c_1(\mathcal{G})}{s} \geq \frac{c_1(\mathcal{F})}{r} \text{ respectively}\right).$$

Remarks : (i) We can assume $\mathcal{G}$ to be torsion-free, since by dividing out a torsion subsheaf, c_1 will not become bigger.

(ii) Another equivalent condition is : For every proper rank-s subsheaf $\mathcal{E} \subseteq \mathcal{F}$ we have $\frac{c_1(\mathcal{E})}{s} < \frac{c_1(\mathcal{F})}{r}$ (or $\frac{c_1(\mathcal{E})}{s} \leq \frac{c_1(\mathcal{F})}{r}$ resp.)

The following proposition can already be found in [11] .

3.1.3. **Proposition.** *If $\mathcal{F}$ is stable then it is simple, and if $c_1(\mathcal{F}) = 0$ then it is s-stable.*

3.1.4. **Proposition.** *If* $\operatorname{rank}(\mathcal{F}) \leq 3$ *then $\mathcal{F}$ is stable if and only if it is s-stable.*

Proof. Assume that there is a torsion-free quotient $\mathcal{G}$ with $c_1(\mathcal{G}) \leq 0$.

Case 1 : rank $\mathcal{G} = 2$. There is an exact sequence

$$0 \longrightarrow \mathcal{E} \longrightarrow \mathcal{F} \longrightarrow \mathcal{G} \longrightarrow 0 .$$

$\mathcal{E}$ is torsion-free, and since $\mathcal{G}$ is torsion-free it is also normal hence an invertible sheaf (cf. [1, Lemma 1]). We conclude $\mathcal{E} = \mathcal{O}(k)$ for some $k \geq 0$ and hence $h^0(\mathcal{F}) \neq 0$.

Case 2 : rank $\mathcal{G} = 1$. If $\operatorname{rank}(\mathcal{F}) = 2$ we can copy the above proof. Otherwise we have an exact sequence

$$0 \longrightarrow \mathcal{G}^* \longrightarrow \mathcal{F}^* \longrightarrow \tilde{\mathcal{G}} \longrightarrow 0$$

with $c_1(\tilde{\mathcal{G}}) = -c_1(\mathcal{G}^*) = c_1(\mathcal{G}) \leq 0$. We set $\hat{\mathcal{G}} := \tilde{\mathcal{G}}/_{T(\tilde{\mathcal{G}})}$ where $T(\tilde{\mathcal{G}})$ is the torsion subsheaf of $\tilde{\mathcal{G}}$. Since $c_1(\hat{\mathcal{G}}) \leq c_1(\tilde{\mathcal{G}})$ we can argue as above to find $h^0(\mathcal{F}^*) \neq 0$.

Remark. (i) For higher rank these notion of stability will in general be different. For example the direct sum of two stable rank-2 bundles will be s-stable again but will not be stable in any other sense.

(ii) For bundles with rank at most three we shall use the word stable without specifying it.

(iii) If $\operatorname{rank}(\mathcal{F}) = 2$ then it suffices to check $h^0(\mathcal{F}) = 0$ for stability since every such bundle carries a symplectic form $\mathcal{F} \cong \mathcal{F}^*$. Here stable is also equivalent to simple [1] .

3.2. In chapter 1 we have proved that the s-stable bundles $\mathcal{F}$ correspond to certain Kronecker-modules $\alpha \in W = \mathrm{Hom}(\mathbb{C}^n \otimes V^*, \mathbb{C}^n)$. On the vector space W the group $GL(n,\mathbb{C}) \times GL(n,\mathbb{C})$ operates as described in (1.2.3.) by

$$\alpha_i \longrightarrow \omega_1 \, \alpha_i \, \omega_2^T \ .$$

As we are interested in stability we want to determine the stable points of W (in the sense of Mumford [10]). To get any stable points at all it is necessary to restrict the operation to $SL(n,\mathbb{C}) \times SL(n,\mathbb{C})$. We then want to compare stability of $\mathcal{F}$ to stability of α . Though our results are still very partial this might indicate a way to get a better hold on the notion of stability which so far is quite difficult to handle.

3.2.1. Definition. A Kronecker-module α is (*properly*) *stable* if for every 1-parameter subgroup $t : \mathbb{C}^* \longrightarrow SL(n,\mathbb{C}) \times SL(n,\mathbb{C})$ we have $\lim\limits_{t \to 0} t(\alpha) = \infty$.
It will be called *unstable* if there is a 1-parameter subgroup t such that $\lim\limits_{t \to 0} t(\alpha) = 0$. If $\mathcal{F}$ is neither unstable nor properly stable it will be called *semi-stable*.

In order to characterize the stable point α we introduce :

3.2.2. Definition. We call α *reducible* if there are subspaces $K, L \subseteq \mathbb{C}^n$ with $0 < \dim K = \dim L < n$ such that $\alpha_i(K) \subseteq L$ for $i = 0,1,2$. Otherwise it will be called *irreducible*. If there are subspaces $K, L \subseteq \mathbb{C}^n$ with $\dim L < \dim K$ and $\alpha_i(K) \subseteq L$ then α will be called *singular*.

Remarks. (i) α is reducible if and only if one can find coordinates and some s with $o < s < n$ such that $(\alpha_i)_{k\ell} = 0$ for $1 \le k \le n-s$ and $1 \le \ell \le s$; i.e. if the matrices α_i can be simultaneously brought to a form

$$\alpha_i = \left[\begin{array}{c|c} 0 & \\ \hline & * \\ * & \end{array}\right]$$

(ii) α is singular if and only if we can find coordinates and some s with $0 < s \leq n$ such that $(\alpha_i)_{k\ell} = 0$ for $1 \leq k \leq n-s+1$ and $1 \leq \ell \leq s$; i.e. if

$$\alpha_i = \left[\begin{array}{c|c} 0 & \\ \hline 0..0 & \ddots \; * \\ \ddots \; * & \end{array}\right]$$

We can now determine the stable Kronecker-modules α . This characterisation was given by C.T.C. Wall [12, Theorem 1] for nets of quadrics.

3.2.3. Proposition. (i) α *is stable if and only if it is irreducible.*

(ii) α *is unstable if and only if it is singular.*

Proof. We first assume that α is not properly stable. Then we can find a 1-parameter subgroup t such that $\lim_{t \to 0} t(\alpha) \neq \infty$. By choosing suitable coordinates we can assume that t is given by diagonal matrices

$$\begin{pmatrix} t^{\gamma_1} & & \\ & \ddots & \\ & & t^{\gamma_n} \end{pmatrix} \qquad \begin{pmatrix} t^{\delta_1} & & \\ & \ddots & \\ & & t^{\delta_n} \end{pmatrix} .$$

where at least one γ_k or δ_ℓ is non-zero. We have $\Sigma\gamma_k = \Sigma\delta_\ell = 0$ and we can assume $\gamma_1 \leq \ldots \leq \gamma_n$ and $\delta_1 \leq \ldots \leq \delta_n$. Then

$$t(\alpha_i)_{k\ell} = (\alpha_i)_{k\ell} \cdot t^{\gamma_k + \delta_\ell} .$$

We claim that there is some i_o with $\gamma_{i_o} + \delta_{n-i_o} < 0$. Namely if this were not the case we would have

$$0 \leq \sum_{k=1}^{n-1} (\gamma_k + \delta_{n-k}) = \sum_{k=1}^{n-1} \gamma_k + \sum_{\ell=1}^{n-1} \delta_\ell .$$

Adding $\gamma_n + \delta_n > 0$ this would imply

$$0 < \sum_{k=1}^{n} \gamma_k + \sum_{\ell=1}^{n} \delta_\ell = 0.$$

Thus we have $\gamma_k + \delta_\ell < 0$ for $k \leq i_o$, $\ell \leq n-i_o$ and it follows from $\lim_{t \to 0} t(\alpha) \neq \infty$ that

$$(\alpha_i)_{k\ell} = 0 \quad \text{for} \quad k \leq i_o \quad \ell \leq n-i_o .$$

Hence α is reducible.

Now assume that α is unstable, i.e. that we may find some t with $\lim_{t \to 0} t(\alpha) = 0$. Again we claim that there is some i_o such that $\gamma_{i_o+1} + \delta_{n-i_o} \leq 0$. Because otherwise we would have

$$0 > \sum_{k=0}^{n-1} \gamma_{k+1} + \delta_{n-k} = \sum_{k=1}^{n} \gamma_k + \sum_{\ell=1}^{n} \delta_\ell = 0 .$$

But now $\lim_{t \to 0} t(\alpha) = 0$ implies

$$(\alpha_i)_{k\ell} = 0 \text{ for } k \leq i_o+1, \quad \ell \leq n-i_o$$

and therefore α is singular.

To prove the converse it is sufficient to find γ_k, δ_ℓ and γ'_k, δ'_ℓ such that for a given s the following holds :

$$\gamma'_k + \delta'_\ell \geq 0 \qquad \text{if } \ell > s \text{ or } k > n-s$$

$$\gamma'_k + \delta'_\ell > 0 \qquad \text{if } \ell > s \text{ or } k > n-s+1.$$

This, however, can be done as follows :

$$\gamma_k := \begin{cases} -s & \text{if } 1 \leq k \leq n-s \\ (n-s) & \text{if } n-s+1 \leq k \leq n \end{cases}$$

$$\delta_\ell := \begin{cases} -(n-s) & \text{if } 1 \leq \ell \leq s \\ s & \text{if } s+1 \leq \ell \leq n \end{cases}$$

and

$$\delta'_\ell := \delta_\ell$$

$$\gamma'_k := \begin{cases} -s+1 & \text{if } 1 \leq k \leq n-s+1 \\ (n-s)+1 & \text{if } n-s+2 \leq k \leq n \end{cases} .$$

Remark. By the above it follows immediately that a stable Kronecker-module is in particular pre-stable.

3.3. We now want to compare stability of the vector bundle $\mathcal{F}$ and stability of the associated Kronecker-module $\alpha_{\mathcal{F}}$. In chapter 1 we have already seen that $\mathcal{F}$ is s-stable if and only if $\alpha_{\mathcal{F}}$ is pre-stable. Our first result says that if α is not properly stable then $\mathcal{F}$ is unstable in the following weak sense :

3.3.1. Proposition. *If $\alpha_{\mathcal{F}}$ is not stable then there exists an s-stable vector bundle $\mathcal{G}$ (with $c_1(\mathcal{G}) = 0$, $c_2(\mathcal{G}) < c_2(\mathcal{F})$) and a non-zero morphism $s : \mathcal{G} \longrightarrow \mathcal{F}$.*

Proof. Since $\alpha = \alpha_{\mathcal{F}}$ is not stable it must be reducible, i.e. we may assume

$$\alpha_i = \left[\begin{array}{c|c} 0 & \\ \hline \alpha_i' & \end{array} \; * \right]$$

where α_i' is a quadratic matrix. If the Kronecker-module given by the α_i' is not pre-stable we can assume

$$\alpha_i' = \begin{pmatrix} 0 & \\ \vdots & \\ 0 & * \\ * & \end{pmatrix} \text{ or } \alpha_i' = \left(\begin{array}{c|c} 0 \,..\, 0 & \\ \alpha_i'' & * \end{array}\right)$$

In the first case this would contradict the pre-stability of α , in the second case we can replace α_i' by α_i''. Proceeding in this way we get a pre-stable Kronecker-module $\beta : \mathbb{C}^m \otimes V^* \longrightarrow \mathbb{C}^n$. This defines a monad.

$$\mathbb{C}^n \otimes \mathcal{O}(-1) \longrightarrow \operatorname{Im} A(\beta) \otimes \mathcal{O} \longrightarrow \mathbb{C}^m \otimes \mathcal{O}(1)$$

and hence an s-stable vector bundle $\mathcal{G}$. Moreover we have a morphism of monads

$$\begin{array}{ccccc} \mathbb{C}^m \otimes \mathcal{O}(-1) & \longrightarrow & \operatorname{Im} A(\beta) \otimes \mathcal{O} & \longrightarrow & \mathbb{C}^m \otimes \mathcal{O}(1) \\ \downarrow & & \downarrow & & \downarrow \\ \mathbb{C}^n \otimes \mathcal{O}(-1) & \longrightarrow & \operatorname{Im} A(\alpha) \otimes \mathcal{O} & \longrightarrow & \mathbb{C}^n \otimes \mathcal{O}(1) \end{array}$$

and hence a map $s : \mathcal{G} \rightarrow \mathcal{F}$. To see that $s \neq 0$ we proceed as follows : We may assume that one of the matrices β_i is non-zero, e.g. $\beta_2 \neq 0$.

Then we choose some $\varphi \in C^m$ with $\beta_2(\varphi) \neq 0$. By $x := A(\beta)(\varphi \otimes v_1) = \beta_2(\varphi) \otimes z_o - \beta_o(\varphi) \otimes z_2$ we define an element in the fibre of $\mathcal{G}$ over (1:0:0) such that $s(x) \neq 0$.

3.3.2. Proposition. *If* $\operatorname{rank}(\mathcal{F})=2$ *then* $\mathcal{F}$ *is stable if and only if* $\alpha_{\mathcal{F}}$ *is stable.*

Proof. If α is stable then clearly $\mathcal{F}$ is stable. To see the converse it is because of (3.1.4.) sufficient to prove that $\alpha_{\mathcal{F}}$ is irreducible. But this follows from [3, Prop. 1.4.].

As an application we get the Grauert-Mülich theorem for stable rank-2 vector bundles with $c_1(\mathcal{F}) = 0$.

3.3.3. Corollary (Grauert-Mülich). *If* $\mathcal{F}$ *is a stable rank-2 vector bundle then* $\Delta(\alpha_{\mathcal{F}}) \not\equiv 0$, *i.e.* $\mathcal{F}|L = \mathcal{O}_L \oplus \mathcal{O}_L$ *for general* L.

Proof. We can assume that the Kronecker-module $\alpha = \alpha_{\mathcal{F}}$ is given by symmetric matrices (see [2]), i.e. that $\alpha \in \operatorname{Hom}(V^*, S^2(\mathbb{C}^n))$. On this vector space the group $SL(n,\mathbb{C})$ operates by $\alpha_i \longrightarrow \omega^T \alpha_i \omega$. Since α was a stable point in $\operatorname{Hom}(\mathbb{C}^n \otimes V^*, \mathbb{C}^n)$ under the operation of $SL(n,\mathbb{C}) \times SL(n,\mathbb{C})$ it follows that not all invariants vanish. But the same will then in particular be true under the operation of $SL(n,\mathbb{C})$. But there the invariants are a finite extension of Δ (see [12, p.232]), and it follows $\Delta(\alpha) \not\equiv 0$.

If $\operatorname{rank}(\mathcal{F}) > 2$ it is unfortunately much more difficult to compare both notions of stability. We only can give the following partial result :

3.3.4. Proposition. *If* $c_2(\mathcal{F}) \leq 7$ *and if* $\operatorname{rank}(\mathcal{F})=3$ *then* $\mathcal{F}$ *is stable if and only if* $\alpha_{\mathcal{F}}$ *is stable.*

Proof. Take some $\mathcal{F}$ and assume that $\alpha_{\mathcal{F}}$ is not stable. Then we can apply the construction of (3.3.1) and find a Kronecker-module $\beta : \mathbb{C}^m \otimes V^* \longrightarrow \mathbb{C}^m$ and an s-stable bundle $\mathcal{G}$ with $0 \neq s : \mathcal{G} \longrightarrow \mathcal{F}$. We first assume that $m \leq 3$. Then $\operatorname{rank}(\mathcal{G}) \leq 3$ and $\mathcal{G}$ is stable. Hence $c_1(s(\mathcal{G})) \geq 0$ which is a contradiction to the stability of $\mathcal{F}$.

Now if $m \geq 4$ we look at $\alpha'^{T}_{\mathscr{F}}$ and applying the same construction we find a subsheaf of $\mathscr{F}^*$ whose first Chern class is non-negative.

3.4. It might be interesting to investigate more closely the relations between the different notions of stability. In any case they will coincide for a (dense) Zariski-open set of bundles. The advantage of considering those bundles for which $\alpha_{\mathscr{F}}$ is stable is that their moduli-scheme could be constructed as a quotient

$\tilde{M}^s(r;0,n)/_{SL(n,\mathbb{C})\times SL(n,\mathbb{C})}$ where $\tilde{M}^s(r;0,n)$ is the quasi-projective variety of stablepoints in $\tilde{M}(r;0,n)$.

REFERENCES

[1] BARTH, W. Some properties of stable rank-2 vector bundles on $\mathbb{P}_n$. Math. Ann. 226, 125-150 (1977).

[2] BARTH, W. Moduli of vector bundles on the projective plane, Invent. math. 42, 63-91 (1977).

[3] BARTH, W., ELENCWAJG, G. Concernant la cohomologie des fibres algébriques stables sur $\mathbb{P}_n(\mathbb{C})$. In Springer LN 683, Berlin-Heidelberg -New-York, Springer 1978.

[4] BARTH, W., HULEK, K. Monads and moduli of vector bundles. Manuscripta Math. 25, 323-347 (1978).

[5] BEILINSON, A.A. Coherent sheaves on $\mathbb{P}^n$ and problems of linear algebra. Funktional'nyi Analiz. i. Ego Prilozheniya, Vol. 12, No. 3, 68-69 (1978).

[6] FULTON, W., HANSEN, J. A connectedness theorem for projective varieties, with applications to intersections and singularities of mappings. To appear in Math. Ann..

[7] MARUYAMA, M. Openness of a family of torsion free sheaves. Journ. of Math. of Kyoto Univ., Vol. 16, Nr. 3, 627-637 (1976).

[8] MARUYAMA, M. Moduli of stable sheaves II. Journ. of Math. of Kyoto Univ., Vol. 18, No.3, 557-614 (1978).

[9] MUMFORD, D. Geometric Invariant Theory. Berlin-Heidelberg- New-York : Springer 1965.

[10] LE POTIER, J. Fibrés stables de rang 2 sur $\mathbb{P}_2(\mathbb{C})$. Math. Ann. 241, 217-256 (1979).

[11] TAKEMOTO, F. Stable vector bundles on algebraic surfaces. Nagoya Math. J. 47, 29-48 (1972).

[12] WALL, C.T.C. Nets of quadrics and theta-characteristics of singular curves. Phil. Transactions of the Royal Soc. of London, A 289, 229-269 (1978).

Klaus HULEK
Mathematisches Institut
der Universität Erlangen-
Nürnberg
Bismarckstr. 1 $\frac{1}{2}$
D-8520 ERLANGEN
WEST-GERMANY.

STABILITÉ ET AMPLITUDE SUR $\mathbb{P}_2(\mathbb{C})$

par

J. LE POTIER

U.E.R. de Mathématiques, Université Paris VII
Tour 45-55, 2 place Jussieu, 75221 PARIS CEDEX 05
France

STABILITÉ ET AMPLITUDE SUR $\mathbb{P}_2(\mathbb{C})$

J.LE POTIER

1. INTRODUCTION

Soit E un fibré algébrique de rang r sur une variété algébrique lisse X. On dit que E est très ample si le fibré J^1E des jets d'ordre un de sections de E est engendré par les sections globales de E. S'il existe un entier n tel que la puissance symétrique S^nE soit très ample, on dit que E est ample. Si $P(E) \xrightarrow{\pi} X$ désigne le fibré en espaces projectifs associé à E.(au-dessus de chaque point $x \in X$, la fibre P(E(x)) est l'espace projectif des hyperplans de E(x)),et $\mathcal{O}_{P(E)}(1)$ le fibré quotient canonique de rang un de π^*E sur P(E), on a l'équivalence [4]

$$E \text{ ample} \Leftrightarrow \mathcal{O}_{P(E)}(1) \text{ ample}$$

Dans le cas où X est une surface de Riemann, et où E est un fibré vectoriel stable, on sait d'après R.Hartshorne que E est ample si et seulement si son degré est > o [7]. Voir qu'un fibré vectoriel stable sur une surface de Riemann est ample ou non est donc une question topologique.

Lorsque $X = \mathbb{P}_2(\mathbb{C})$, on connaît des conditions topologiques nécessaires pour qu'un fibré E soit ample : si on identifie les classes de Chern de E à des nombres c_1 et $c_2 \in \mathbb{Z}$ grâce aux isomorphismes canoniques $H^{2i}(\mathbb{P}_2,\mathbb{Z}) \simeq \mathbb{Z}$ pour i = 1 et 2, ces conditions nécessaires s'écrivent [8]

$$c_1 > o \;,\quad c_2 > o \quad, c_1^2 - c_2 > o \qquad \textbf{(A)}$$

Même pour les fibrés stables de rang 2 sur $\mathbb{P}_2$, ces conditions ne sont pas suffisantes pour assurer l'amplitude. Rappelons tout d'abord que si c_1 et $c_2 \in \mathbb{Z}$ sont donnés, il existe un fibré stable de rang 2 de classes de Chern c_1 et c_2 si et seulement si [12]

$$\text{et} \quad \begin{cases} 4\,c_2 - c_1^2 > o \\ 4\,c_2 - c_1^2 \neq 4 \end{cases} \qquad \textbf{(B)}$$

Si les conditions **(A)** et **(B)** sont réalisées, on a automatiquement $c_1 \geq 2$.

Pour $c_1 = 2$, elles imposent $c_2 = 3$; il est facile de voir qu'aucun fibré stable de rang 2 de classes de Chern (2,3) sur $\mathbb{P}_2$ n'est ample (corollaire **3.2**). Pour $c_1 = 3$, ces conditions signifient $3 \leq c_2 \leq 8$; cependant, on peut vérifier que le seul fibré stable de rang 2 de classe de Chern $c_2 = 3$ qui soit ample est le fibré tangent $T(\mathbb{P}_2)$ (corollaire **3.3**).

Pourtant, M.Schneider trouve dans [11] des conditions suffisantes qui permettent d'assurer qu'un fibré stable de rang 2 sur $\mathbb{P}_2$ est très ample: ces conditions s'écrivent :

$$c_2 - \frac{c_1^2}{4} \leq \frac{c_1}{2}$$

La question se pose donc de savoir si l'amplitude est encore une notion topologique pour les fibrés stables sur $\mathbb{P}_2$, c'est-à-dire si elle peut se décrire en termes d'inégalités portant sur les classes de Chern. Nous montrons ici qu'il n'en est rien. On examine pour ceci les fibrés stables de rang 2 de classe de Chern $c_1 = 4$; les conditions **(A)** et **(B)** ci-dessus imposent $6 \leq c_2 \leq 15$. Le cas $c_2 = 6$ est réglé par le critère de Schneider: ils sont tous très amples. Nous ferons une étude détaillée des cas $c_2 = 7$, $c_2 = 8$.

On désigne par $M(c_1, c_2)$ la variété des classes d'isomorphismes de fibrés vectoriels stables de rang 2 de classes de Chern c_1 et c_2 ; c'est une variété algébrique lisse et connexe de dimension $4c_2 - c_1^2 - 3$. Il n'existe pas toujours de fibré universel sur $M(c_1, c_2) \times \mathbb{P}_2$; cependant, nous désignons pour $s \in M(c_1, c_2)$ par $E(s)$ un fibré représentant la classe d'isomorphisme s.
Soit $\mathrm{Pic}\, M(c_1,c_2)$ le groupe de Picard des classes d'isomorphismes de fibrés vectoriels algébriques de rang 1 sur $M = M(c_1,c_2)$. Si c_1 est pair, on a la description suivante [10] :

$$\mathrm{Pic}\, M(c_1,c_2) = \begin{cases} \mathbb{Z} & \text{si } c_2 - \frac{c_1^2}{4} > 2 \\ \mathbb{Z}/3\mathbb{Z} & \text{si } c_2 - \frac{c_1^2}{4} = 2 \end{cases}$$

Il est possible de faire un choix entre les 2 générateurs de $\mathrm{Pic}\, M(c_1,c_2)$: en effet, si E est un fibré de rang 2 sur $\mathbb{P}_2$, de classes de Chern c_1 et c_2 , avec c_1 pair, les droites de saut de E, c'est-à-dire les droites $\ell \subset \mathbb{P}_2$ sur lesquelles E n'est pas isomorphe à $\mathcal{O}\left(\frac{c_1}{2}\right) \oplus \mathcal{O}\left(\frac{c_1}{2}\right)$ forment une courbe $\sigma(E)$ de degré $c_2 - \frac{c_1^2}{4}$ dans le plan projectif $\mathbb{P}_2^\vee$ dual; ceci détermine un morphisme

$$\sigma : M(c_1,c_2) \to \mathbb{P}_N(\mathbb{C})$$

dans l'espace projectif des courbes de degré $c_2 - \frac{c_1^2}{4}$ dans $\mathbb{P}_2^\vee$. On peut vérifier (cf. paragraphe **2.3**) que

- si $c_2 - \frac{c_1^2}{4}$ est impair, $\sigma^*\left(\mathcal{O}_{P_N}(1)\right) = L_o$ est un des deux générateurs de Pic $M(c_1,c_2)$,
- si $c_2 - \frac{c_1^2}{4}$ est pair, $\sigma^*\left(\mathcal{O}_{P_N}(1)\right) = L_o^{\otimes 2}$ avec L_o générateur de Pic $M(c_1,c_2)$.

Ce choix étant fait, on pourra parler du degré d'une hypersurface $\Sigma \subset M(c_1,c_2)$, ce qui permet d'énoncer les résultats concernant M(4,7) et M(4,8).

Théorème 1.1 a) Pour tout $s \in M(4,7)$ le fibré E(s) est engendré par ses sections globales.

b) L'ensemble des points $s \in M(4,7)$ où E(s) n'est pas ample est une hypersurface lisse Σ de degré 4. En dehors de Σ, E(s) est très ample.

Théorème 1.2 a) L'ensemble des points $s \in M(4,8)$ où E(s) n'est pas ample est une hypersurface Σ_o de degré 15 ; en dehors de Σ_o , E(s) est très ample.

b) L'ensemble des points $s \in M(4,8)$ où E(s) n'est pas engendré par ses sections globales est une sous-variété lisse Σ_1 de codimension 3 contenue dans le lieu singulier $S(\Sigma_o)$ de Σ_o .

On peut vérifier que pour $c_1 = 4$, les seuls fibrés stables très amples de rang 2 sont ceux pour lesquels $c_2 = 6$, et ceux qui sont obtenus dans ces deux énoncés (proposition **3.6**). Ainsi, pour ces fibrés, $c_2 \leqslant 8$. Si l'on pose

$$\chi = \chi(c_1,c_2) = 2 + \frac{c_1(c_1+3)}{2} - c_2$$

cette condition s'écrit aussi $\chi(c_1,c_2) \geqslant 8$. L'énoncé suivant généralise en quelque sorte ce résultat.

Théorème 1.3 Il existe dans $M(c_1,c_2)$ un ouvert de Zariski non vide correspondant à des fibrés très amples si et seulement si, outre les conditions **(B)** on a

$$c_1 > o \quad \underline{\text{et}} \quad \chi(c_1,c_2) \geqslant 8 \ .$$

On peut naturellement se demander si la condition $\chi(c_1,c_2) \geqslant 8$ est nécessaire pour qu'un fibré stable de rang 2 soit très ample. Il n'en est rien: il existe des fibrés très amples tels que $\chi(c_1,c_2) < o$ (remarque **6.15**). Au voisinage du point s_o défini par un tel fibré E_o , E(s) ne reste pas très ample; cependant, il reste ample (corollaire **7.3**). Ceci permet de démontrer l'existence de fibrés amples sur P_2 qui n'ont aucune section globale non identiquement nulle.

Les points $s \in M(c_1,c_2)$ tels que $E(s)$ soit ample forment un ouvert de Zariski. Concernant cet ouvert, la question suivante n'a reçu qu'une réponse partielle.

QUESTION : A quelles conditions l'ouvert de Zariski des points $s \in M(c_1,c_2)$ tels que $E(s)$ soit ample est-il non vide ?

De manière plus précise, nous obtenons au chapitre 7 une réponse affirmative si $c_1 \geqslant 4$ et $\chi \geqslant 4$. Par exemple, les fibrés obtenus comme conoyau d'un morphisme injectif

$$\mathcal{O}(-1)^4 \xrightarrow{f} \mathcal{O}^6$$

forment un ouvert de Zariski dans $M(4,10)$ dont aucun n'est très ample; dans cette famille, il existe un ouvert correspondant à des fibrés amples.

NOTATIONS : Si F est un faisceau algébrique cohérent sur une variété algébrique X de faisceau structural $\mathcal{O}$, on désigne par F_x le $\mathcal{O}_x$-module des germes de sections de F au point x, et par $F(x)$ le $\mathbb{C}$-espace vectoriel

$$F(x) = F_x \otimes_{\mathcal{O}_x} \mathbb{C}$$

Soit m_x l'idéal maximal de $\mathcal{O}_x$; l'espace $J^1F(x) = F_x \otimes_{\mathcal{O}_x} \mathcal{O}_x/m_x^2$ s'identifie, lorsque F est localement libre, à l'espace vectoriel des jets d'ordre 1 de sections de F au point x.

2. GENERALITES

2.1 La suite spectrale de Beilinson. On utilise l'essentiel des notations de [9]. En particulier, on désigne par Q le fibré quotient canonique de rang 2 de $\mathcal{O}^3$ sur $\mathbb{P}_2$, et on pose $V = \Gamma(\mathbb{P}_2, \mathcal{O}(Q))$; le dual V^* s'identifie à $\Gamma(\mathbb{P}_2, \mathcal{O}(1))$.

Soit E un fibré vectoriel algébrique sur $\mathbb{P}_2$. On sait d'après [3] qu'il existe une suite spectrale (dans la catégorie des $\mathcal{O}$-modules cohérents sur $\mathbb{P}_2$) dont le terme $E_1^{p,q}$ est donné par

$$E_1^{p,q} = \Lambda^{-p} Q^* \otimes H^q(\mathbb{P}_2, E(p))$$

et qui a pour aboutissement $\left\{ \begin{array}{l} E \text{ en degré } 0 \\ 0 \text{ en degré } \neq 0 \end{array} \right.$

Dans cette suite spectrale, le morphisme $d_1^{p,q} : E_1^{p,q} \longrightarrow E_1^{p+1,q}$ est donné par le morphisme de fibrés associé à l'accouplement naturel $z \otimes h \longrightarrow z \cdot h$

$$V^* \otimes H^q(\mathbb{P}_2, E(p)) \longrightarrow H^q(\mathbb{P}_2, E(p+1))$$

de la manière suivante : pour $p = -2$, on a au-dessus du point $x \in \mathbb{P}_2$

$$d_1^{-2,q}(z' \wedge z'' \otimes h) = z'' \otimes z' \cdot h - z' \otimes z'' \cdot h$$

pour $z', z'' \in Q_x^*$, $h \in H^q(\mathbb{P}_2, E(-2))$. Pour $p = -1$, $z \in Q_x^*$, $h \in H^q(\mathbb{P}_2, E(-1))$ on a

$$d_1^{-1,q}(z \otimes h) = z \cdot h$$

De plus, le morphisme canonique

$$E_1^{o,o} = \mathcal{O} \otimes H^o(\mathbb{P}_2, E) \longrightarrow E$$

associé à cette suite spectrale n'est autre que le morphisme d'évaluation noté ev dans la suite.

2.2 Cas des fibrés stables. Dans le cas où E est un fibré stable de rang 2 de classe de Chern $c_1 = o$, on a $H^o(\mathbb{P}_2, E(i)) = o$ pour $i \leqslant o$, et $H^2(\mathbb{P}_2, E(i)) = o$ pour $i \geqslant -3$. Ceci permet d'obtenir E comme fibré de cohomologie de la monade [9]

$$\wedge^2 Q^* \otimes H \xrightarrow{d_1^{-2,1}} Q^* \otimes H^* \xrightarrow{d_1^{-1,1}} \mathcal{O} \otimes K$$

où $H = H^1(\mathbb{P}_2, E(-2))$, $K = H^1(\mathbb{P}_2, E)$; on a $\dim H = c_2$ et $\dim K = c_2 - 2$.

Réciproquement, si H est un espace vectoriel de dimension c_2, K est un espace vectoriel de dimension $c_2 - 2$, S^2H^* l'espace vectoriel des applications linéaires symétriques $H \to H^*$, à tout couple (α, β) d'applications linéaires

$$\alpha : V^* \to S^2H^*$$
$$\beta : V^* \to L(H^*, K)$$

on associe des morphismes de fibrés vectoriels

$$\wedge^2 Q^* \otimes H \xrightarrow{a} Q^* \otimes H^* \xrightarrow{b} \mathcal{O} \otimes K .$$

Soit P l'espace des couples (α, β) pour lesquels la suite ci-dessus est une monade, c'est-à-dire telle que a soit injectif, b surjectif, et $b \circ a = o$; soit $E(\alpha, \beta)$ le fibré de cohomologie associé: c'est un fibré stable de rang 2 de classes de Chern (o, c_2). L'espace P est une variété algébrique lisse sur laquelle le groupe $G = GL(H) \times GL(K)/_{\{\pm 1\}}$ opère librement par la formule

$$((f,g),(\alpha,\beta)) \longmapsto \left({}^t f^{-1} \alpha f^{-1},\ g \beta\, {}^t f\right)$$

pour $f \in GL(H)$, $g \in GL(K)$, $(\alpha, \beta) \in P$. Le quotient P/G muni de la topologie quotient et du faisceau des fonctions régulières inversibles, est une variété algébrique lisse quasi-projective, qui s'identifie à $M(o, c_2)$. Sur l'espace $P \times \mathbb{P}_2$, on construit de manière évidente un fibré universel noté **E** dans la suite.

On peut pour $c_1 = -1$ donner une description analogue de $M(-1, c_2)$ [9]. Il en résulte que $M(c_1,c_2)$ est irréductible [2,13] .

2.3 Description de Pic $M(o,c_2)$ [10]. On désigne par Pic $M(o,c_2)$ le groupe des classes d'isomorphisme de fibrés vectoriels algébriques de rang 1 sur $M = M(o,c_2)$, et par $G^\vee$ le groupe des caractères de G. A tout caractère $\chi \in G^\vee$, on associe une structure de G-fibré vectoriel L'_χ sur le fibré trivial $P \times \mathbb{C}$; le quotient $L_\chi = L'_\chi/G$ est alors un G-fibré vectoriel de rang 1 sur $M(o,c_2)$. On a ainsi défini un homomorphisme de groupes

$$\lambda : G^\vee \longrightarrow \text{Pic } M(o,c_2)$$

Proposition 2.1 : <u>Le morphisme de groupe défini ci-dessus</u>

$$\lambda : G^\vee \longrightarrow \text{Pic } M(o,c_2)$$

<u>est un épimorphisme dont le noyau est le sous-groupe engendré par le caractère</u> $\chi_o \in G^\vee$ <u>défini par</u>

$$\chi_o(f,g) = (\det f)^3 \det g$$

<u>pour</u> $f \in GL(H)$, $g \in GL(K)$.

Démonstration. Considérons l'inclusion $GL(K) \subset G$ définie par $g \longmapsto (1,g)$ mod.$\{\pm 1\}$. Sur le quotient $P_1 = P/_{GL(K)}$ le groupe $G_1 = G/_{GL(K)} \simeq GL(H)/_{\{\pm 1\}}$ opère librement, et le quotient P_1/G_1 est encore isomorphe à $M(o,c_2)$. On a encore un morphisme $\lambda_1 : G_1^\vee \longrightarrow \text{Pic } M(o,c_2)$ et le diagramme

$$\begin{array}{ccc} G_1^\vee & \xrightarrow{\lambda_1} & \\ \downarrow & & \text{Pic } M(o,c_2) \\ G^\vee & \xrightarrow{\lambda} & \end{array}$$

est commutatif. D'après [10], paragraphe 3, le morphisme λ_1 est un isomorphisme sauf pour $c_2 = 2$, auquel cas c'est un épimorphisme. Par suite λ est un épimorphisme. Le groupe des caractères $G^\vee$ est donné, suivant les valeurs de c_2, par le tableau suivant :

	$G^\vee$	générateurs
$c_2 = 2$	$\mathbb{Z}$	$f \longmapsto \det f$
c_2 pair > 2	$\mathbb{Z} \oplus \mathbb{Z}$	$(f,g) \longmapsto \det f$; $(f,g) \longmapsto \det g$
c_2 impair	$\mathbb{Z} \oplus \mathbb{Z}$	$(f,g) \longmapsto (\det f)^2$; $(f,g) \longmapsto \det f \det g$

Au caractère $\chi_o : (f,g) \longmapsto (\det f)^3 \det g$ correspond le fibré $(\det E)^*/G|M \times \{x_o\}$, où x_o est un point de $\mathbb{P}_2$. On sait d'après [10], paragraphe 5, que ce fibré est trivial. Ainsi, $\chi_o \in \text{Ker } \lambda$, et on obtient un épimorphisme

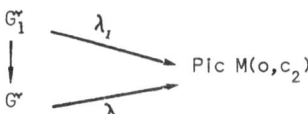

$$G^*/_{(\chi_o)} \longrightarrow \text{Pic } M(o,c_2)$$

Il est clair que $G^*/_{(\chi_o)} \simeq \begin{cases} Z/_{3Z} & \text{si } c_2 = 2 \\ Z & \text{si } c_2 > 2 \end{cases}$, et on la la même

description pour Pic $M(o,c_2)$ d'après [10]. Par suite, l'épimorphisme ci-dessus est un isomorphisme, et Ker $\lambda = (\chi_o)$.

Proposition 2.2 : Le fibré $\sigma^*\left(\mathcal{O}_{P_N}(1)\right)$ est le fibré associé au caractère $(f,g) \longmapsto (\det f)^{-2}$ de G.

La démonstration est donnée dans [10], paragraphe 3. Ceci entraîne en particulier que si L_o est le générateur de Pic $M(o,c_2)$ associé au caractère

$$(f,g) \longmapsto \begin{cases} (\det f)^{-1} & \text{si } c_2 \text{ est pair} \\ (\det f)^{-2} & \text{si } c_2 \text{ est impair} \end{cases}$$

on aura

$$\sigma^*\left(\mathcal{O}_{P_N}(1)\right) \simeq \begin{cases} L_o & \text{si } c_2 \text{ est impair} \\ L_o^{\otimes 2} & \text{si } c_2 \text{ est pair} \end{cases}$$

3. MAJORATION DE dim $H^o(\mathbb{P}_2,E)$

Soit E un fibré vectoriel stable de rang 2 sur $\mathbb{P}_2$, de classes de Chern c_1 et c_2 ; l'objet de ce paragraphe est de préciser quelles sont les valeurs possibles de c_2 si E est ample et $2 \leqslant c_1 \leqslant 4$.

Rappelons tout d'abord que si E est stable, on a $H^o(\mathbb{P}_2,E) = o$ dès que $c_1 \leqslant o$. Ceci implique par dualité de Serre $H^2(\mathbb{P}_2,E) = o$ pour $c_1 \geqslant -6$. D'autre part, d'après le théorème de Grauert-Mülich [1], sur une droite générique $\ell \subset \mathbb{P}_2$, on peut écrire

$$E|_\ell \simeq \mathcal{O}(m) \oplus \mathcal{O}(n)$$

avec m et $n \in \mathbf{Z}$ et $|m-n| = o$ ou 1 . La suite exacte longue de cohomologie appliquée à ℓ montre que si $c_1 \geqslant -4$, dim $H^1(\mathbb{P}_2,E(1)) \leqslant$ dim $H^1(\mathbb{P}_2,E)$. La proposition suivante précise ce résultat.

Proposition 3.1 : Soit E un fibré stable de rang 2 de classes de Chern c_1 et c_2 sur $\mathbb{P}_2$:

a) Si $c_1 \geq -3$ et $H^1(\mathbb{P}_2,E) \neq 0$, on a

$$\dim H^1(\mathbb{P}_2,E(1)) < \dim H^1(\mathbb{P}_2,E)$$

b) Si E(3) est ample et $c_1 \geq -4$, on a

$$\dim H^1(\mathbb{P}_2,E(1)) < \dim H^1(\mathbb{P}_2,E) - 1$$

sauf si $\dim H^1(\mathbb{P}_2,E) = 0$ ou 1.

Démonstration. a) Dans les cas $c_1 = -3$ ou -2, on sait calculer les dimensions de $H^1(\mathbb{P}_2,E)$ et $H^1(\mathbb{P}_2,E(1))$ par la formule de Riemann-Roch : le résultat est donné par le tableau suivant :

c_1	-3	-2
$\dim H^1(\mathbb{P}_2,E)$	$c_2 - 2$	$c_2 - 1$
$\dim H^1(\mathbb{P}_2,E(1))$	$c_2 - 1$	$c_2 - 3$

Dans le cas $c_1 \geq -1$, on sait d'après M.Schneider [11] que si $\dim H^1(\mathbb{P}_2,E) = \dim H^1(\mathbb{P}_2,E(1))$, on a sur toute droite $\ell \subset \mathbb{P}_2$ $E(1)|_\ell \simeq \mathcal{O}(m) \oplus \mathcal{O}(n)$, avec m et $n \in \mathbb{Z}$, $m \geq 0$ et $n \geq 0$. Considérons sur le plan projectif dual $\mathbb{P}_2^\vee$ des droites de $\mathbb{P}_2$ le morphisme

$$H^1(\mathbb{P}_2,E) \otimes \mathcal{O}_{\mathbb{P}_2^\vee}(-1) \xrightarrow{A} H^1(\mathbb{P}_2,E(1)) \otimes \mathcal{O}_{\mathbb{P}_2^\vee}$$

défini par $h \otimes z \longmapsto z \cdot h$ pour $h \in H^1(\mathbb{P}_2,E)$, $z \in V^*$.
Au-dessus du point $\ell \in \mathbb{P}_2^\vee$, le conoyau de ce morphisme est $H^1(\ell,E(1))$; il devrait donc être surjectif, et par suite un isomorphisme. C'est impossible, sauf si les deux membres sont nuls.

b) Supposons $c_1 \geq -3$. Si

$$\dim H^1(\mathbb{P}_2,E(1)) = \dim H^1(\mathbb{P}_2,E) - 1$$

le morphisme A n'est pas surjectif, sinon le noyau de A serait un fibré de rang un de classe de Chern $1 - du + \frac{1}{2}d(d-1)u^2$, où $d = \dim H^1(\mathbb{P}_2,E)$ et $u = c_1(\mathcal{O}_{\mathbb{P}_2^\vee}(1))$ ce qui est absurde pour $d > 1$. Ceci signifie qu'il existe des droites $\ell \subset \mathbb{P}_2$ telles que $E|_\ell \simeq \mathcal{O}(m) \oplus \mathcal{O}(c_1 - m)$, avec $m \leq -3$. Par suite E(3) a un quotient de rang un de classe de Chern ≤ 0, et donc E(3) n'est pas ample.

Dans le cas $c_1 = -4$, on a $\dim H^1(\mathbb{P}_2,E(1)) = \dim H^1(\mathbb{P}_2,E)$ et il y a encore des droites vérifiant la condition ci-dessus; on a donc le même résultat.

Corollaire 3.2 : Il n'y a pas de fibré stable ample de rang 2 sur $\mathbb{P}_2$ tel que $c_1 = 2$.

Démonstration. Soit E un fibré stable de rang 2 sur $\mathbb{P}_2$ de classe de Chern $c_1 = 2$. On a $\dim H^1(\mathbb{P}_2, E(-2)) = \dim H^1(\mathbb{P}_2, E(-3)) \geqslant 2$; par conséquent, le corollaire résulte de la proposition **3.1**, assertion **b)**.

Corollaire 3.3 : Le seul fibré stable ample de rang 2 sur $\mathbb{P}_2$ de classe de Chern $c_1 = 3$ est le fibré tangent $T(\mathbb{P}_2)$.

Démonstration. Soit E un fibré stable de rang 2 de classe de Chern $c_1 = 3$. Alors $c_1(E(-2)) = -1$, et par suite

$$\dim H^1(\mathbb{P}_2, E(-2)) = \dim H^1(\mathbb{P}_2, E(-3)) - 1 = c_2 - 3$$

Il résulte de la proposition **3.1** que si $c_2 \neq 3$, E n'est pas ample. L'espace M(3,3) étant réduit à un seul point donné par le fibré tangent $T(\mathbb{P}_2)$, ceci démontre le corollaire **3.3**.

On pose $\quad \chi = \chi(c_1, c_2) = 2 + \dfrac{c_1(c_1+3)}{2} - c_2$.

Proposition 3.4 : Soit E un fibré stable de rang 2 sur $\mathbb{P}_2$ de classes de Chern c_1 et c_2 ; on suppose $c_1 \geqslant 0$.

1°) On a

$$\dim H^0(\mathbb{P}_2, E) \leqslant \begin{cases} \operatorname{Max}\left(\chi, \dfrac{c_1(c_1+4)}{4}\right) & \text{si } c_1 \text{ est pair} \\[2ex] \operatorname{Max}\left(\chi, \dfrac{(c_1+1)(c_1+3)}{4}\right) & \text{si } c_1 \text{ est impair} \end{cases}$$

2°) En particulier, si $\dfrac{c_1^2}{4} < c_2 \leqslant \dfrac{c_1^2}{4} + \dfrac{c_1}{2} + 2$

$$\dim H^0(\mathbb{P}_2, E) = \chi \quad \text{et} \quad H^1(\mathbb{P}_2, E) = 0$$

3°) Si E est ample de classe de Chern $c_1 \geqslant 4$, on a

$$\dim H^0(\mathbb{P}_2, E) \leqslant \begin{cases} \operatorname{Max}\left(\chi, \dfrac{c_1(c_1+4)}{4} - 2\right) & \text{si } c_1 \text{ est pair} \\[2ex] \operatorname{Max}\left(\chi, \dfrac{(c_1+1)(c_1+3)}{4} - 2\right) & \text{si } c_1 \text{ est impair} \end{cases}$$

Démonstration. Supposons c_1 pair. Alors

$$\dim H^1\left(\mathbb{P}_2, E(-\frac{c_1}{2})\right) = c_2 - \frac{c_1^2}{4} - 2$$

Il résulte de la proposition **3.1** que l'on a

$$\dim H^1(\mathbb{P}_2,E) \begin{cases} \leqslant c_2 - \frac{c_1^2}{4} - \frac{c_1}{2} - 2 & \text{si } 4c_2 \geqslant c_1^2 + 2c_1 + 8 \\ = 0 & \text{si } 4c_2 \leqslant c_1^2 + 2c_1 + 8 \end{cases}$$

Dans le premier cas, la formule de Riemann-Roch entraîne

$$\chi \leqslant \dim H^0(\mathbb{P}_2,E) \leqslant \frac{c_1(c_1+4)}{4}$$

Supposons c_1 impair. Alors

$$\dim H^1\left(\mathbb{P}_2, E(-\frac{c_1+1}{2})\right) = c_2 - \frac{c_1^2-1}{4} - 1$$

La proposition **3.1** donne la majoration

$$\dim H^1(\mathbb{P}_2,E) \begin{cases} \leqslant c_2 - \frac{c_1^2 + 2c_1 + 5}{4} & \text{si } 4c_2 \geqslant c_1^2 + 2c_1 + 5 \\ = 0 & \text{si } 4c_2 \leqslant c_1^2 + 2c_1 + 5 \end{cases}$$

Dans le premier cas, la formule de Riemann-Roch conduit encore à la formule désirée. Compte-tenu du fait que

$$\chi - \frac{c_1(c_1+4)}{4} = 2 + \frac{c_1^2}{4} + \frac{c_1}{2} - c_2$$

ceci donne les assertions 1°) et 2°). La partie 3°) s'obtient en remarquant que si E est ample et $\dim H^1(\mathbb{P}_2,E) \neq 0$, on a d'après la proposition **3.1**

$$\dim H^1(\mathbb{P}_2,E) \leqslant \dim H^1(\mathbb{P}_2,E(-2)) - 4$$

Lemme 3.5 : <u>Soit</u> E <u>un fibré vectoriel sur</u> $\mathbb{P}_2$. <u>Si l'on pose</u> $u = c_1(\mathcal{O}(1))$, <u>la classe de Chern totale du fibré des jets</u> J^1E <u>est donnée par</u>

$$c(J^1E) = c(E(-1))^3 = (1 + (c_1 - 2)u + (c_2 - c_1 + 1)u^2)^3$$

Démonstration. On a la suite exacte de fibrés vectoriels sur $\mathbb{P}_2$

$$0 \longrightarrow T^* \otimes E \longrightarrow J^1E \longrightarrow E \longrightarrow 0$$

D'autre part, la suite exacte $0 \longrightarrow T^* \longrightarrow \mathcal{O}(-1)^3 \to \mathcal{O} \to 0$ donne par tensorisation avec E la suite exacte

$$0 \longrightarrow T^* \otimes E \longrightarrow E(-1)^3 \longrightarrow E \longrightarrow 0$$

Il en résulte que J^1E et $E(-1)^3$ ont même classe de Chern d'où le lemme.

Proposition 3.6 : Soit E un fibré stable très ample de rang 2, de classes de Chern $c_1 = 4$ et c_2 sur $\mathbb{P}_2$; alors $c_2 = 6,7$ ou 8.

Démonstration. Si E est un fibré stable ample de rang 2 de classe de Chern $c_1 = 4$ sur $\mathbb{P}_2$, on a $\chi + c_2 = 16$, et d'après les conditions **(A)** et **(B)** (cf.**Introduction**), $6 \leqslant c_2 \leqslant 15$. D'après la proposition **3.4**, on obtient

$$\dim H^\circ(\mathbb{P}_2,E) \begin{cases} = \chi & \text{si } c_2 = 6,7 \text{ ou } 8 \\ \leqslant 7 & \text{si } c_2 = 9 \\ \leqslant 6 & \text{si } 10 \leqslant c_2 \leqslant 15 \end{cases}$$

Montrons que pour $c_2 > 8$, le morphisme canonique

$$H^\circ(\mathbb{P}_2,E) \otimes \mathcal{O} \longrightarrow J^1E$$

ne peut être surjectif. Si $c_2 = 9$, $\dim H^\circ(\mathbb{P}_2,E) = 7$, si ce morphisme était surjectif, le noyau serait un fibré de rang 1, L, de classe de Chern

$$c(L) = \frac{1}{c(J^1E)} = 1 - 6u + 6u^2$$

ce qui est absurde. De même, le morphisme ci-dessus ne peut être un isomorphisme dans le cas $\dim H^\circ(\mathbb{P}_2,E) = 6$, car $c_1(J^1E) = 6$. Pour $\dim H^\circ(\mathbb{P}_2,E) < 6$, c'est une question de rang. Par suite, les seules valeurs possibles pour c_2 si E est très ample sont $c_2 = 6,7$ ou 8.

Remarque 3.7. 1°) Tous les fibrés stables de rang 2 de classes de Chern $c_1 = 4$, $c_1 = 6$ sont très amples d'après le critère de Schneider [11] .

2°) On verra au paragraphe **7** que pour $c_1 = 4$, $6 \leqslant c_2 \leqslant 12$, il existe des fibrés stables amples de classes de Chern c_1 et c_2 .

3°) J'ignore s'il existe des fibrés stables amples de rang 2 tels que $c_1 = 4$, $13 \leqslant c_2 \leqslant 15$.

4. ETUDE DE M(4,7)

Ce chapitre est consacré à la démonstration du théorème **1.1.** Si E est un fibré stable de rang 2 de classes de Chern (4,7) sur $\mathbb{P}_2$, on a

$$\dim H^1(\mathbb{P}_2,E(-3)) = 3 \ ; \ \dim H^1(\mathbb{P}_2,E(-2)) = 1$$

Proposition 4.1 : Soit E un fibré stable de classe de Chern $c_1 = 4$ et $c_2 = 7$, de rang 2 sur $\mathbb{P}_2$. Alors $H^1(\mathbb{P}_2, E(-1)) = o$ (et par suite E est engendré par ses sections globales). De plus, les conditions suivantes sont équivalentes :
(1) E est très ample, **(2)** E est ample, **(3)** Le morphisme canonique sur $\mathbb{P}_2^\vee$

$$B : H^1(\mathbb{P}_2, E(-3)) \otimes \mathcal{O}_{\mathbb{P}_2^\vee}(-1) \longrightarrow H^1(\mathbb{P}_2, E(-2)) \otimes \mathcal{O}_{\mathbb{P}_2^\vee}$$

est surjectif.

Démonstration. Du fait que $\dim H^1(\mathbb{P}_2, E(-2)) = 1$, on a bien d'après la proposition **3.1** $H^1(\mathbb{P}_2, E(-1)) = o$. Si m_x est l'idéal du point $x \in \mathbb{P}_2$, on aura $H^1(\mathbb{P}_2, m_x E) = o$,ce qui entraîne que E est engendré par ses sections globales.

On a toujours (1) ⇒ (2), et (2) ⇒ (3) en répétant l'argument utilisé dans la démonstration de la proposition **3.1.** Il reste à vérifier que (3) ⇒ (1).

Lemme 4.2 Soit E un fibré de rang 2, de classes de Chern (4,7), tel que le morphisme canonique sur $\mathbb{P}_2^\vee$

$$B : H^1(\mathbb{P}_2, E(-3)) \otimes \mathcal{O}_{\mathbb{P}_2^\vee}(-1) \longrightarrow H^1(\mathbb{P}_2, E(-2)) \otimes \mathcal{O}_{\mathbb{P}_2^\vee}$$

soit surjectif. Alors E est le conoyau d'un morphisme injectif de fibrés

$$\mathcal{O}(-1) \longrightarrow \mathcal{O}(1)^3 .$$

Démonstration. Posons pour $z \in V^*$, $h \in H^1(\mathbb{P}_2, E(-3))$, $B(z)h = B(h \otimes z)$, et considérons la suite spectrale de Beilinson pour E(-1). Le morphisme $d_1^{-2,1} : E_1^{-2,1} \to E_1^{-1,1}$:

$$\Lambda^2 Q^* \otimes H^1(\mathbb{P}_2, E(-3)) \longrightarrow Q^* \otimes H^1(\mathbb{P}_2, E(-2))$$

est donné au-dessus du point $x \in \mathbb{P}_2$ par $z' \wedge z'' \otimes h \longmapsto B(z')h \otimes z'' - B(z'')h \otimes z'$ pour $h \in H^1(\mathbb{P}_2, E(-3))$ et z', $z'' \in Q_x^*$. Si z' et z'' sont indépendants, on aura au-dessus de x :

$$\operatorname{Ker} d_1^{-2,1} = \operatorname{Ker} B(z') \cap \operatorname{Ker} B(z'')$$

L'hypothèse entraîne que les sous-espaces Ker B(z') et Ker B(z") sont de dimension 2 et distincts dans un espace de dimension 3, par suite, Ker $d_1^{-2,1}$ est de dimension 1 au-dessus de chaque point, et $d_1^{-2,1}$ est de rang 2. Le noyau $Z = \operatorname{Ker} d_1^{-2,1}$ est donc un fibré de rang 1, de classe de Chern

$$c_1(Z) = (1-u)^3(1+u) = 1 - 2u$$

où l'on a posé $u = c_1(\mathcal{O}(1))$. Par conséquent $Z \simeq \mathcal{O}(-2)$. La suite spectrale ci-dessus donne une suite exacte de $\mathcal{O}$- modules (et donc de fibrés)

$$\begin{array}{ccccccc} 0 \longrightarrow & E_2^{-2,1} & \xrightarrow{d_2^{-2,1}} & E_2^{0,0} & \longrightarrow & E(-1) & \longrightarrow 0 \\ & \wr\!\wr & & \wr\!\wr & & & \\ & \mathcal{O}(-2) & & \mathcal{O}^3 & & & \end{array}$$

d'où découle le lemme **4.2.**

Démonstration du théorème 1.1. L'assertion **a)** a déjà été vue; démontrons **b)**. Considérons la projection décrite au paragraphe **2.2**

$$P \longrightarrow M(o,3) \simeq M(4,7)$$

qui permet d'obtenir sur M(4,7) une structure de variété algébrique lisse de dimension 9, et posons, pour $(\alpha,\beta) \in P$, $E = E(\alpha,\beta)(2)$, de telle sorte que le morphisme B sur $\mathbb{P}_2$ associé à E considéré ci-dessus s'identifie au morphisme

$$H^* \otimes \mathcal{O}_{\mathbb{P}_2^\vee}(-1) \longrightarrow K \otimes \mathcal{O}_{\mathbb{P}_2^\vee}$$

associé à l'application linéaire β: $V^* \longrightarrow L(H^*,K)$. L'ensemble Σ des points $s \in M(4,7)$ où E(s) n'est pas ample est donc le lieu des zéros de la section det β du fibré de rang 1 sur M(4,7) défini par

$$L = P \times \mathrm{Hom}(\Lambda^3 V^*, \det H \otimes K^{\otimes 3})/G$$

D'après la proposition **2.1**, ce fibré est isomorphe à $P \times (\det H)^{\otimes -8}/G \simeq L_o^{\otimes 4}$. Il reste à montrer que la section obtenue est transverse à la section nulle. Puisque l'application linéaire $\beta : V^* \longrightarrow L(H^*,K)$ est de rang $\geqslant 2$ ([2] § 7) il suffit de vérifier le lemme suivant

Lemme 4.3. : <u>La projection</u> $(\alpha,\beta) \longmapsto \beta$:

$$P \longrightarrow L(V^*, L(H^*,K))$$

<u>est une submersion</u>.

Démonstration. On sait [9] que l'espace tangent à P au point (α,β) est donné par les couples $(u,v) \in \mathrm{Hom}(\Lambda^2 Q^* \otimes H, Q^* \otimes H^*) \times \mathrm{Hom}(Q^* \otimes H^*, \mathcal{O} \otimes K)$ vérifiant les conditions suivantes

- u provient d'une application linéaire $V^* \longrightarrow S^2H^*$
- $v \circ a + b \circ u = o$, où a et b sont les morphismes figurant dans la monade associée au couple (α,β) (cf.paragraphe 2.2).

L'application linéaire tangente à la projection ci-dessus est donnée par $(u,v) \longmapsto v$. Pour voir qu'elle est surjective, on remarque tout d'abord que v étant donné, le morphisme $-v \circ a$ se factorise suivant le diagramme

$$\begin{array}{ccc} \Lambda^2 Q^* \otimes H & & \\ \Big\downarrow w & \searrow{\scriptstyle -v\circ a} & \\ Q^* \otimes H^* & \xrightarrow{\ b\ } & \mathcal{O}\otimes K \end{array}$$

En effet b est surjectif, et d'après la proposition **4.1**, $H^1(\mathbb{P}_2, \mathrm{Ker}\, b\,(1))$ est nul. D'autre part, il résulte de [9], lemme 31, que le morphisme $w : \Lambda^2 Q^* \otimes H \to Q^* \otimes H^*$ peut s'écrire

$$w = u - a \circ f$$

où f est un endomorphisme de $\Lambda^2 Q^* \otimes H$, et $u : \Lambda^2 Q^* \otimes H \longrightarrow Q^* \otimes H^*$ un morphisme provenant d'une application linéaire $V^* \to S^2 H^*$. De la relation $b \circ w + v \circ a = o$ on tire $b \circ u + v \circ a = o$, ce qui montre que $(u,v) \in T_{(\alpha,\beta)}P$. D'où le lemme 4.3.

Remarque 4.4. Si $c_1 = o$ et $c_2 > 3$, la démonstration ci-dessus montre que l'ouvert des couples $(\alpha,\beta) \in P$ tels que $H^1(\mathbb{P}_2, E(\alpha,\beta)(1)) = o$, la projection $(\alpha,\beta) \longmapsto \beta : P \longrightarrow L(V^*, L(H^*, K))$ est une submersion. On verra au chapitre **6** que cet ouvert est non vide si et seulement si $2 \leqslant c_2 \leqslant 6$.

5. ETUDE DE M(4,8)

On se propose dans ce chapitre de démontrer le théorème **1.2**. Si E est un fibré stable de rang 2 sur $\mathbb{P}_2$ de classes de Chern (4,8), on a

$$\dim H^1(\mathbb{P}_2, E(-3)) = 4 \quad ; \quad \dim H^1(\mathbb{P}_2, E(-2)) = 2$$

et par suite $\dim H^1(\mathbb{P}_2, E(-1)) = o$ ou 1.

Proposition 5.1 Soit E un fibré stable de rang 2 sur $\mathbb{P}_2$, de classes de Chern (4,8). Les assertions suivantes sont équivalentes.

(1) E est engendré par ses sections globales

(2) $H^1(\mathbb{P}_2, E(-1)) = o$

Démonstration. (1) ⇒ (2) : si $\dim H^1(\mathbb{P}_2, E(-1)) = 1$, il existe une droite $\ell \subset \mathbb{P}_2$ telle que $E(-1)|_\ell \simeq \mathcal{O}(-2) \oplus \mathcal{O}(4)$, et donc E n'est pas engendré par ses sections globales.

(2) ⇒ (1) : pour tout $x \in \mathbb{P}_2$, on a la suite exacte

$$0 \longrightarrow \mathcal{O}(-2) \longrightarrow \mathcal{O}(-1)^2 \longrightarrow m_x \longrightarrow 0 .$$

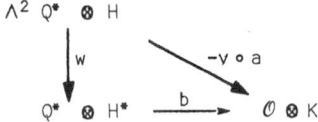

La suite exacte longue de cohomologie montre que $H^1(\mathbb{P}_2, m_x E) = o$, et par suite E est engendré par ses sections globales.

Proposition 5.2 : Les points $s \in M(4,8)$ tels que le fibré $E(s)$ ne soit pas engendré par ses sections globales forment une sous-variété algébrique fermée Σ_1 lisse de codimension 3.

La démonstration de cette proposition repose sur le lemme suivant, qui donne une description des fibrés qui ne sont pas engendrés par leurs sections.

Lemme 5.3. : Soit E un fibré stable de rang 2, de classes de Chern (4,8) sur $\mathbb{P}_2$. Les assertions suivantes sont équivalentes:

(1) $\dim H^1(\mathbb{P}_2, E(-1)) = 1$

(2) il existe une droite $\ell \subset \mathbb{P}_2$ et une suite exacte de $\mathcal{O}$-modules

$$0 \longrightarrow T \longrightarrow E \longrightarrow \mathcal{O}_\ell(-1) \longrightarrow 0$$

où T est le fibré tangent à $\mathbb{P}_2$.

Démonstration. Il est evident que **(2)** ⇒ **(1)** puisque $H^1(\mathbb{P}_2, T(-1)) = o$. Montrons que **(1)** ⇒ **(2)**. La suite spectrale de Beilinson appliquée à E(-1) montre que le morphisme canonique $Q^* \otimes H^1(\mathbb{P}_2, E(-2)) \longrightarrow \mathcal{O} \otimes H^1(\mathbb{P}_2, E(-1))$ est surjectif. Il en résulte que sur $\mathbb{P}_2^\vee$, le morphisme

$$H^1(\mathbb{P}_2, E(-2)) \otimes \mathcal{O}_{\mathbb{P}_2^\vee}(-1) \longrightarrow H^1(\mathbb{P}_2, E(-1)) \otimes \mathcal{O}_{\mathbb{P}_2^\vee}$$

s'annule en un seul point $\ell \in \mathbb{P}_2^\vee$. Par suite, le morphisme $d_1^{-1,2}$ de la suite spectrale de Beilinson de E

$$\Lambda^2 Q^* \otimes H^1(\mathbb{P}_2, E(-2)) \longrightarrow Q^* \otimes H^1(\mathbb{P}_2, E(-1))$$

est de rang 2 en dehors de $\ell \subset \mathbb{P}_2$ et de rang 1 sur ℓ. Le $\mathcal{O}$-module $\mathcal{G} = \operatorname{coker} d_1^{-1,2}$ est un $\mathcal{O}_\ell$-module localement libre de rang 1; on constate que $H^1(\mathbb{P}_2, \mathcal{G}(-1)) \simeq H^1(\ell, \mathcal{G}(-1))$ est de dimension 1, d'où il résulte que $\mathcal{G} \simeq \mathcal{O}_\ell(-1)$. Or, on a la suite exacte de $\mathcal{O}$-modules

$$0 \longrightarrow Q^* \otimes H^\circ(\mathbb{P}_2, E(-1)) \xrightarrow{d_1^{-1,0}} \mathcal{O} \otimes H^\circ(\mathbb{P}_2, E) \xrightarrow{ev} E \longrightarrow \mathcal{G} \longrightarrow 0$$

Il résulte du fait que $\operatorname{Tor}_2^{\mathcal{O}_x}(\mathcal{G}_x, \mathbb{C}) = o$ pour tout $x \in \mathbb{P}_2$ que le morphisme $d_1^{-1,0}$ est injectif comme morphisme de fibrés. Le conoyau $T = \operatorname{coker} d_1^{-1,0}$ est un fibré de rang 2 de classes de Chern (3,3); on vérifie de plus que $H^\circ(\mathbb{P}_2, T(-2)) = o$, d'où il résulte que T est stable. Par conséquent, T est isomorphe au fibré tangent à $\mathbb{P}_2$.

Lemme 5.4. : Il existe des fibrés stables de rang 2 sur $\mathbb{P}_2$ de classes de Chern (4,8), E, tels que $H^1(\mathbb{P}_2,E(-1)) \neq o$.

Démonstration. Soit ℓ une droite de $\mathbb{P}_2$. Le groupe des extensions

$$0 \longrightarrow T \longrightarrow E \longrightarrow \mathcal{O}_\ell(1) \longrightarrow 0$$

est classé par le groupe $\mathrm{Ext}^1(\mathcal{O}_\ell(-1),T)$, qui est un $\mathbb{C}$-espace vectoriel de dimension 9; un tel module E sera localement libre si pour tout $x \in \ell$ $\mathrm{Tor}_1^{\mathcal{O}_x}(E_x,\mathbb{C}) = o$, c'est-à-dire si l'application linéaire associée à l'extension ci-dessus

$$\mathrm{Tor}_1^{\mathcal{O}_x}(\mathcal{O}_\ell(-1)_x,\mathbb{C}) \longrightarrow T(x)$$

est injective. La résolution $0 \longrightarrow \mathcal{O}(-2) \xrightarrow{z} \mathcal{O}(-1) \longrightarrow \mathcal{O}_\ell(-1) \longrightarrow 0$ conduit à la suite exacte

$$0 \longrightarrow \Gamma(\mathbb{P}_2,T(1)) \xrightarrow{z} \Gamma(\mathbb{P}_2,T(2)) \longrightarrow \mathrm{Ext}^1(\mathcal{O}_\ell(-1),T) \longrightarrow 0$$

Si s est une section de T(2), l'extension associée à E sera localement libre si et seulement si le morphisme de fibré $\mathcal{O}(-2) \xrightarrow{(z,s)} \mathcal{O}(-1) \oplus T$ est injectif, ce qui revient à dire que s est non nulle sur ℓ. Le fibré T(2) étant très ample, on peut toujours trouver une telle section s. On vérifie que le fibré associé est stable de rang 2 de classes de Chern (4,8); de la suite exacte définissant E comme extension découle que $\dim H^1(\mathbb{P}_2,E(-1)) = 1$.

Démonstration de la proposition 5.2. Considérons la projection décrite au paragraphe 2.2 $\pi : P \to M(o,4) \simeq M(4,8)$ qui permet d'obtenir sur M(4,8) une structure de variété algébrique lisse de dimension 13. Soit Σ_1 l'ensemble des points de M(4,8) correspondant à des fibrés qui ne sont pas engendrés par leurs sections globales. Il suffit bien entendu de vérifier que l'image réciproque P_1 de Σ_1 par la projection π est une sous-variété fermée lisse de codimension 3.

Pour $(\alpha,\beta) \in P$, désignons par E le fibré $E(\alpha,\beta)(2)$. Les groupes de cohomologie $H^q(\mathbb{P}_2,E(-1))$ sont alors les groupes de cohomologie du complexe associé à (α,β) :

$$\begin{array}{ccccccc} o \longrightarrow & \Gamma(\mathbb{P}_2,\mathcal{O}\otimes H) & \longrightarrow & \Gamma(\mathbb{P}_2,Q\otimes H^*) & \longrightarrow & \Gamma(\mathbb{P}_2,\Lambda^2 Q\otimes K) & \longrightarrow 0 \\ & \| & & \| & & \| & \\ & H & & V\otimes H^* & & \Lambda^2 V\otimes K & \end{array}$$

D'après le théorème de semi-continuité, l'ensemble des couples $(\alpha,\beta) \in P$ tels que $H^1(\mathbb{P}_2,E(-1)) = o$ est un fermé P_1. Si $(\alpha,\beta) \in P_1$, au voisinage W de (α,β), le complexe ci-dessus, correspondant au fibré $E' = E(\alpha',\beta')(2)$, est quasi-isomorphe à un complexe :

$$0 \longrightarrow H^o(\mathbb{P}_2, E(-1)) \xrightarrow{\beta_1'} H^1(\mathbb{P}_2, E(-1)) \longrightarrow 0$$

où β_1' dépend algébriquement de (α',β'). Le morphisme $(\alpha',\beta') \longmapsto \beta_1'$ s'annule au point (α,β), et sa différentielle au point (α,β) est indépendante du quasi-isomorphisme choisi; elle se factorise suivant le diagramme

$$\begin{array}{ccc} T_{(\alpha,\beta)}P & & \\ \downarrow & \searrow & \\ H^1(\mathbb{P}_2, \underline{Hom}(E,E)) & \longrightarrow & L(H^o(\mathbb{P}_2, E(-1)), H^1(\mathbb{P}_2 E(-1))) \end{array}$$

où la flèche horizontale est l'accouplement canonique. Le fermé P_1 est donné sur W par l'équation $\beta_1' = o$; la proposition sera donc démontrée si on vérifie que le morphisme $(\alpha',\beta') \longmapsto \beta_1'$ est transverse à 0 au point $(\alpha,\beta) \in P_1$, c'est-à-dire que la différentielle ci-dessus est surjective. D'après le lemme 5.3, on a une suite exacte $0 \longrightarrow T \longrightarrow E \longrightarrow \mathcal{O}_\ell(-1) \longrightarrow 0$, où ℓ est une droite de $\mathbb{P}_2$ qui induit les isomorphismes

$$H^o(\mathbb{P}_2, E(-1)) \simeq H^o(\mathbb{P}_2, T(-1)) \simeq H^o(\ell, T(-1))$$

$$H^1(\mathbb{P}_2, E(-2)) \simeq H^1(\ell, \mathcal{O}_\ell(-2)) .$$

et un épimorphisme

$$H^1(\mathbb{P}_2, \underline{Hom}(E,E)) \longrightarrow H^1(\ell, T^*(-1)) .$$

On est donc ramené à voir que l'accouplement naturel

$$H^1(\ell, T^*(-1)) \longrightarrow L(H^o(\ell, T(-1)), H^1(\ell, \mathcal{O}_\ell(-2)))$$

est un isomorphisme: ceci résulte du théorème de dualité de Serre pour la droite ℓ.

Proposition 5.5 : Soit E un fibré stable de rang 2 sur $\mathbb{P}_2$ de classes de Chern (4,8). Les assertions suivantes sont équivalentes

(1) E est très ample

(2) E est ample

(3) le morphisme canonique sur $\mathbb{P}_2^\vee$

$$B : H^1(\mathbb{P}_2, E(-3)) \otimes \mathcal{O}_{\mathbb{P}_2^\vee}(-1) \longrightarrow H^1(\mathbb{P}_2, E(-2)) \otimes \mathcal{O}_{\mathbb{P}_2^\vee}$$

est surjectif.

Démonstration. Les implications (1) ⇒ (2) et (2) ⇒ (3) se voient comme dans la démonstration de la proposition **4.1.** L'implication **(3) ⇒ (1)** résulte du lemme suivant:

Lemme 5.6.: On suppose B surjectif. Dans la suite spectrale de Beilinson de E(-1), le morphisme

$$d_1^{-2,1}: \Lambda^2 Q^* \otimes H^1(\mathbb{P}_2, E(-3)) \longrightarrow Q^* \otimes H^1(\mathbb{P}_2, E(-2))$$

est un morphisme de $\mathcal{O}$-modules injectif dont le conoyau $\mathcal{G}$ est isomorphe à $\mathcal{O}_C(-a)$, où C est une conique lisse de $\mathbb{P}_2$, et a un point de C.

Démonstration. Soit Z le fibré de rang 2 sur $\check{\mathbb{P}}_2$ noyau de l'épimorphisme B ; ce fibré a pour classes de Chern (-4,6), et on a $H^2(\check{\mathbb{P}}_2, Z(-1)) = 0$, ce qui s'écrit encore par dualité de Serre $H^\circ(\check{\mathbb{P}}_2, Z(2)) = 0$; ceci signifie que Z est un fibré stable. Soit ℓ_x la droite de $\check{\mathbb{P}}_2$ définie par le point $x \in \mathbb{P}_2$.

On a alors $Z|_{\ell_x} \simeq \mathcal{O}(-2) \oplus \mathcal{O}(-2)$, sauf si x appartient à une conique lisse C de $\mathbb{P}_2$, auquel cas $Z|_{\ell_x} \simeq \mathcal{O}(-3) \oplus \mathcal{O}(-1)$. Plus précisément, désignons par $D \subset \mathbb{P}_2 \times \check{\mathbb{P}}_2$ la variété des couples (x,ℓ) tels que $x \in \ell$, et par $p: D \longrightarrow \mathbb{P}_2$ et $q: D \longrightarrow \check{\mathbb{P}}_2$ les projections canoniques. Alors ([2], paragraphe 2.1.)

(i) $p_*(q^*(Z(1))) = 0$

(ii) le $\mathcal{O}_{\mathbb{P}_2}$ - module $R^1 p_*(q^*(Z(1)))$ est un $\mathcal{O}_C$-module localement libre de rang 1.

Or, la flèche $d_1^{-2,1}$ s'identifie à $p_* q^*(B(1))(-1)$. L'assertion **(i)** signifie que $d_1^{-2,1}$ est injectif comme morphisme de $\mathcal{O}_{\mathbb{P}_2}$-modules; l'assertion **(ii)** signifie que $\mathcal{G}$ est un $\mathcal{O}_C$-module localement libre de rang 1. De plus on a $H^\circ(\mathbb{P}_2, \mathcal{G}) = H^1(\mathbb{P}_2, \mathcal{G}) = 0$, ce qui montre que $\mathcal{G} \simeq \mathcal{O}_C(-a)$, avec $a \in C$. Ceci démontre le lemme **5.6.**

Fin de la démonstration de la proposition 5.5. Si B est surjectif, la suite spectrale de Beilinson de E(-1) montre que $H^1(\mathbb{P}_2, E(-1)) = 0$ et que le morphisme $ev: \mathcal{O} \otimes H^\circ(\mathbb{P}_2, E(-1) \longrightarrow E(-1)$ a même noyau et conoyau que $d_1^{-2,1}$; par suite, on obtient la suite exacte de $\mathcal{O}$-modules

$$0 \longrightarrow \mathcal{O} \otimes H^\circ(\mathbb{P}_2, E(-1)) \xrightarrow{ev} E(-1) \longrightarrow \mathcal{G} \longrightarrow 0$$

ce qui donne une suite exacte $0 \longrightarrow \mathcal{O}(1)^2 \longrightarrow E \longrightarrow \mathcal{G}(1) \longrightarrow 0$. Soit $x \in \mathbb{P}_2, m_x$ l'idéal maximal de $\mathcal{O}_x$. On a le diagramme commutatif de suites exactes :

$$\begin{array}{ccccccc}
0 \longrightarrow & H^\circ(\mathbb{P}_2, \mathcal{O}(1))^2 & \longrightarrow & H^\circ(\mathbb{P}_2, E) & \longrightarrow & H^\circ(\mathbb{P}_2, \mathcal{G}(1)) & \longrightarrow 0 \\
& \downarrow & & \downarrow & & \downarrow & \\
& J^1\mathcal{O}(1)^2(x) & \longrightarrow & J^1 E(x) & \longrightarrow & \mathcal{G}(1)_x \otimes_{\mathcal{O}_x} \mathcal{O}_x/m_x^2 & \longrightarrow 0
\end{array}$$

Puisque $\mathcal{O}(1)$ est très ample, la première flèche verticale est surjective. D'autre part, si $x \in C$, la dernière flèche verticale s'identifie au morphisme canonique :

$$H^{o}(C,\mathcal{G}(1)) \longrightarrow J^{1}(\mathcal{G}(1)|C)(x).$$

Le fibré sur C défini par $\mathcal{G}(1)|C$ étant très ample, on voit que cette flèche est toujours surjective. Par suite, la deuxième flèche verticale est surjective, et donc E est très ample.

Remarque 5.6. Puisque $H^1(\mathbb{P}_2,E) = o$, dire que E est très ample revient à dire que $H^1(\mathbb{P}_2,m_x^2 E) = o$ pour tout $x \in \mathbb{P}_2$. Ceci peut être vérifié directement lorsque B est surjectif en remarquant que $H^1(\mathbb{P}_2,m_x^2 E)$ est le conoyau du morphisme $(H^1(\mathbb{P}_2,E(-3)))^2 \longrightarrow (H^1(\mathbb{P}_2,E(-2)))^3$ donné par la matrice

$$\begin{pmatrix} B(z_2) & 0 \\ -B(z_1) & -B(z_2) \\ 0 & B(z_1) \end{pmatrix}$$

où z_1 et z_2 sont les équations de deux droites distinctes passant par x.

Proposition 5.7.: <u>L'ensemble Σ_o des points $s \in M(4,8)$ correspondant à des fibrés qui ne sont pas amples est une hypersurface de degré 15 dont le lieu singulier contient Σ_1.</u>

La démonstration de cette proposition repose sur deux lemmes qui permettent de décrire Σ_o comme lieu des zéros d'une section non nulle d'un fibré de rang 1 sur M(4,8).

Lemme 5.8.: <u>On a</u> $\Sigma_o \neq M(4,8)$.

Démonstration. Il suffit de vérifier que si C est une conique non dégénérée de $\mathbb{P}_2$, a un point de C, et $\mathcal{L}_a$ le $\mathcal{O}_C$-module défini par le point a, il existe une extension

$$0 \longrightarrow \mathcal{O}(1)^2 \longrightarrow E \longrightarrow \mathcal{L}_a \longrightarrow 0$$

où E est un $\mathcal{O}$-module localement libre; cette extension définit un fibré de rang 2, stable, de classes de Chern (4,8), très ample d'après l'argument utilisé dans la démonstration de la proposition **5.5**. L'existence d'une telle extension se vérifie comme dans le lemme **5.4**.

Lemme 5.9.: <u>L'ensemble Σ_o est l'ensemble des points $s \in M(4,8)$ pour lesquels la forme quadratique $\delta = \det d_1^{-1,2}$ associée au morphisme $d_1^{-2,1}$ du lemme **5.6** est dégénérée.</u>

Démonstration. On a déjà vu que si B est surjective, la forme quadratique δ, qui est l'équation de la conique C du lemme 5.6, est non dégénérée. Réciproquement, s'il existe $z \in V^*$ tel que $B(z)$ soit de rang $\leqslant 1$, en tout point x de la droite ℓ d'équation $z = o$, on aura $\operatorname{Ker} d_1^{-2,1}(x) \neq o$, et par conséquent δ s'annule sur ℓ. Ainsi δ est dégénérée.

Démonstration de la proposition 5.7. Dans la projection $\pi : P \longrightarrow M(4,8)$ l'image réciproque de cette forme quadratique est donnée par une application notée encore $\delta : P \longrightarrow \det H \otimes (\det K)^{\otimes 2} \otimes S^2V^*$, invariante par l'action naturelle du groupe $GL(H) \times GL(K)$. Le discriminant Δ de cette forme quadratique est une section G-invariante du G-fibré de rang 1

$$L = P \times (\det H)^{\otimes 3} \otimes (\det K)^{\otimes 6} \otimes (\Lambda^3 V^*)^{\otimes 2}$$

non identiquement nulle, où $G = GL(H) \times GL(K)/_{\{\pm 1\}}$; compte tenu de la proposition 2.1, on a encore $L \simeq P \times (\det H)^{\otimes -15}$. Il en résulte que

$$L/G \simeq L_o^{\otimes 15}$$

et donc Σ_o est une hypersurface de degré 15.

Ouvert de lissité de Σ_o. Soit P_o l'image réciproque de Σ_o par π. L'ouvert de lissité de Σ_o correspond aux couples $(\alpha,\beta) \in P_o$ tels que

- $\delta(\alpha,\beta)$ soit de rang 2
- l'application $(\alpha',\beta') \longmapsto \delta(\alpha',\beta')$ soit transverse en (α,β) à la variété des formes quadratiques $q \in \det H \otimes (\det K)^{\otimes 2} \otimes S^2V^*$ de rang 2. Cette sous-variété a pour espace normal au point q l'espace $S^2(\operatorname{Ker} q)^*$ des formes quadratiques sur Ker q. La condition de transversalité sera réalisée en un point $(\alpha,\beta) \in P_o$ où $\delta(\alpha,\beta)$ est de rang 2 si et seulement si pour $e \in \operatorname{Ker} \delta(\alpha,\beta)$ non nul, l'application

$$(\alpha',\beta') \longmapsto \delta(\alpha',\beta')(e)$$

a une différentielle non nulle en (α,β). Ceci impose que $d_1^{-2,1}$ soit de rang 3 au point $x \in \mathbb{P}_2$ défini par le noyau de $\delta = \delta(\alpha,\beta) = \det d_1^{-2,1}$. La projection $(\alpha,\beta) \longmapsto \beta$ étant une submersion sur l'ouvert $P - P_1$ d'après la remarque 4.4, cette condition sera suffisante pour assurer la transversalité sur $P - P_1$.

Si $(\alpha,\beta) \in P_1$, on peut trouver $z \in V^* - \{o\}$ tel que $B(z) = o$. La droite ℓ d'équation $z = o$ est alors contenue dans la conique C d'équation $\delta = o$, et le morphisme $d_1^{-2,1}$ est de rang $\leqslant 2$ en tout point de ℓ, et par suite au centre de cette conique (dans l'hypothèse où δ serait de rang 2). Par suite Σ_1 est contenu dans le lieu singulier $S(\Sigma_o)$ de Σ_o. Ceci démontre la proposition **5.7.**

Le théorème **1.2** résulte des propositions **5.2** et **5.7.**

6. FIBRÉS GÉNÉRIQUES

Soit E un fibré stable de rang 2 sur $\mathbb{P}_2$, de classes de Chern c_1 et c_2. Rappelons que si $c_1 \geqslant -6$, on a $H^2(\mathbb{P}_2,E) = o$, et que la caractéristique d'Euler-Poincaré de E est donnée par

$$\chi = \chi(c_1,c_2) = 2 + \frac{c_1(c_1+3)}{2} - c_2$$

On rappelle en outre que $M(c_1,c_2)$ est une variété irréductible. Le but de ce chapitre est d'étudier les propriétés des fibrés stables génériques, c'est-à-dire les propriétés qui sont satisfaites par tous les fibrés d'un ouvert de Zariski de $M(c_1,c_2)$; on démontrera en particulier le théorème **1.3.**

Proposition 6.1. : 1°) <u>Si</u> $\chi(c_1,c_2) \leqslant o$, <u>on a sur un ouvert de Zariski non vide de</u> $M(c_1,c_2)$

$$H^o(\mathbb{P}_2,E) = o$$

2°) <u>Si</u> $c_1 \geqslant o$ <u>et</u> $\chi(c_1,c_2) \geqslant o$, <u>on a sur un ouvert de Zariski non vide de</u> $M(c_1,c_2)$

$$H^1(\mathbb{P}_2,E) = o \quad .$$

Démonstration. Le théorème de semi-continuité, appliqué au fibré universel sur $P \times \mathbb{P}_2$ du paragraphe **2.2,** montre que les conditions

$$H^o(\mathbb{P}_2,E) = o$$
$$H^1(\mathbb{P}_2,E) = o$$

sont ouvertes dans $M(c_1,c_2)$; il reste donc à vérifier que les ouverts correspondants sont non vides. Posons pour $i \in \mathbb{Z}$

$$\chi(i) = \chi(E(i)) = \chi(c_1+2i, c_2+c_1 i + i^2)$$

On a $\chi(i+1) - \chi(i) = c_1+2i+4$; la suite $i \longmapsto \chi(i)$ est donc croissante pour $c_1+2i \geqslant -4$. D'autre part, sa valeur minimum

$$\begin{cases} -\left(c_2 - \dfrac{c_1^2}{4}\right) & \text{si } c_1 \text{ est pair} \\ -\left(c_2 - \dfrac{c_1^2 - 1}{4}\right) & \text{si } c_1 \text{ est impair} \end{cases}$$

est négative. Par conséquent, il existe un plus petit entier k tel que $\chi(k) > o$

et $c_1 + 2k > -4$, et on aura alors $\chi(k) \leqslant c_1 + 2k + 2$. Cet entier k est $> o$ si $\chi(c_1, c_2) \leqslant o$, $\leqslant o$ si $\chi(c_1, c_2) > o$. Quitte à remplacer $M(c_1, c_2)$ par $M(c_1, 2k, c_2 + kc_1 + k^2)$, on est ramené à démontrer la proposition suivante :

Proposition 6.2.: Supposons $c_1 > o$ et $o < \chi(c_1,c_2) \leqslant c_1+2$. L'ouvert de Zariski W de $M(c_1,c_2)$ correspondant aux fibrés E vérifiant les conditions suivantes

$$H^o(\mathbb{P}_2, E(-1)) = o \quad ; \quad H^1(\mathbb{P}_2, E) = o$$

est non vide.

Ceci implique que pour tout entier $i \geqslant o$, $H^o(\mathbb{P}_2, E(-i-1)) = o$ et $H^1(\mathbb{P}_2, E(i)) = o$, ce qui donne la proposition 6.1. Avant de démontrer la proposition 6.2, on se propose de montrer comment la suite spectrale de Beilinson permet d'obtenir pour certains fibrés de W une présentation particulièrement simple.

Soit E un fibré de W; on a alors $\dim H^1(\mathbb{P}_2, E(-2)) = 2c_1 + 2 - \chi$, $\dim H^1(\mathbb{P}_2, E(-1)) = c_1 + 2 - \chi$, et $\dim H^o(\mathbb{P}_2, E) = \chi$. La suite spectrale de Beilinson pour E se réduit à deux complexes

$$0 \longrightarrow \mathcal{O}(-1)^{2c_1+2-\chi} \xrightarrow{d_1^{-2,1}} (Q^*)^{c_1+2-\chi} \longrightarrow 0 \qquad (D_1^\bullet)$$

$$0 \longrightarrow \mathcal{O}^{\chi} \xrightarrow{\quad ev \quad} E \longrightarrow 0 \qquad (D_2^\bullet)$$

qui ont mêmes groupes de cohomologie Z et $\mathcal{G}$. Considérons la suite spectrale d'aboutissement $\mathrm{Ext}^{p+q}(D_1^\bullet, D_2^\bullet)$, de terme E_2

$$E_2^{p,q} = H^p(H^q(\mathbb{P}_2, \underline{\mathrm{Hom}}^\bullet(D_1^\bullet, D_2^\bullet)))$$

où $\underline{\mathrm{Hom}}^\bullet(D_1^\bullet, D_2^\bullet)$ est le complexe des fibrés d'homomorphismes de $D_1^\bullet$ dans $D_2^\bullet$. Compte-tenu du fait que $H^1(\mathbb{P}_2, E(i)) = o$ pour $i \geqslant o$ on aura pour tout $q > o$ $H^q(\mathbb{P}_2, \underline{\mathrm{Hom}}^p(D_1^\bullet, D_2^\bullet)) = o$. Il en résulte un isomorphisme

$$H^p(\mathrm{Hom}^\bullet(D_1^\bullet, D_2^\bullet)) \simeq \mathrm{Ext}^p(D_1^\bullet, D_2^\bullet)$$

où $\mathrm{Hom}^\bullet(D_1^\bullet, D_2^\bullet)$ est le complexe des homomorphismes de $D_1^\bullet$ dans $D_2^\bullet$. D'autre part, on a aussi une suite spectrale de terme E_2

$$'E_2^{p,q} = \prod_i \mathrm{Ext}^p(H^i(D_1^\bullet), H^{i+q}(D_2^\bullet))$$

de même aboutissement, ce qui donne la suite exacte

$$\mathrm{Ext}^o(D_1^\bullet, D_2^\bullet) \longrightarrow \mathrm{Hom}(Z,Z) \oplus \mathrm{Hom}(\mathcal{G},\mathcal{G}) \longrightarrow \mathrm{Ext}^2(\mathcal{G}, Z)$$

Désignons par θ_E l'image de $(id_Z, id_{\mathcal{G}})$ dans $\mathrm{Ext}^2(\mathcal{G}, Z)$, et considérons la partie W_1 de W correspondant aux fibrés E tels que $\theta_E = o$.

Lemme 6.3.: Tout fibré correspondant à un point de W_1 est conoyau d'un morphisme injectif

$$\varphi : \mathcal{O}(-1)^{2c_1+2-\chi} \longrightarrow (Q^*)^{c_1+2-\chi} \oplus \mathcal{O}^{\chi}$$

En particulier, tout fibré de W engendré par ses sections globales est conoyau d'un tel morphisme injectif.

Démonstration. L'hypothèse $\theta_E = o$ signifie qu'il existe un morphisme de complexes $D_1^\bullet \longrightarrow D_2^\bullet$ qui induit l'identité sur Z et $\mathcal{G}$:

$$\begin{array}{ccc} \mathcal{O}(-1)^{2c_1+2-\chi} & \longrightarrow & (Q^*)^{c_1+2-\chi} \\ \downarrow & & \downarrow \\ \mathcal{O}^{\chi} & \longrightarrow & E \end{array}$$

Ceci permet de construire un complexe $K^\bullet$(c'est le "mapping-cône" du morphisme ci-dessus)

$$0 \longrightarrow \mathcal{O}(-1)^{2c_1+2-\chi} \xrightarrow{\varphi} \mathcal{O}^{\chi} \oplus (Q^*)^{c_1+2-\chi} \longrightarrow E \longrightarrow 0$$

tel que l'on ait la suite exacte de complexe

$$\begin{array}{ccccccccc} & & & & 0 & & 0 & & \\ & & & & \downarrow & & \downarrow & & \\ & & 0 & \longrightarrow & \mathcal{O}(-1)^{2c_1+2-\chi} & = & \mathcal{O}(-1)^{2c_1+2-\chi} & \longrightarrow & 0 \\ & & \downarrow & & \downarrow \varphi & & \downarrow & & \\ 0 & \longrightarrow & \mathcal{O}^{\chi} & \longrightarrow & \mathcal{O}^{\chi} \oplus (Q^*)^{c_1+2-\chi} & \longrightarrow & (Q^*)^{c_1+2-\chi} & \longrightarrow & 0 \\ & & \downarrow & & \downarrow & & \downarrow & & \\ 0 & \longrightarrow & E & = & E & \longrightarrow & 0 & & \\ & & \downarrow & & \downarrow & & & & \\ & & 0 & & 0 & & & & \end{array}$$

Il en résulte que le complexe $K^\bullet$ est acyclique; puisque E est localement libre, le flèche φ est obligatoirement injective comme morphisme de fibrés,ce qui démontre le lemme **6.3.**

Lemme 6.4.: Soit φ un morphisme de fibrés vectoriels

$$\mathcal{O}(-1)^{2c_1+2-\chi} \longrightarrow (Q^*)^{c_1+2-\chi} \oplus \mathcal{O}^{\chi}$$

Si φ est générique, alors φ est injectif (comme morphisme de fibrés).

Démonstration. Le fibré sur $\mathbb{P}_2$ $\underline{\mathrm{Hom}}(\mathcal{O}(-1)^{2c_1+2-\chi}, (Q^*)^{c_1+2-\chi} \oplus \mathcal{O}^{\chi})$ est évidemment engendré par ses sections globales. Soient Γ l'espace vectoriel de ses sections globales, Y le sous-ensemble algébrique de $\mathbb{P}_2 \times \Gamma$ des couples (x,φ) tels que $\varphi(x)$ ne soit pas injective. Ce sous-ensemble algébrique Y est un fibré au-dessus de $\mathbb{P}_2$ en sous-ensembles algébriques de codimension 3 dans Γ. Par suite $\dim Y = \dim \Gamma - 1$. Il en résulte que la projection $Y \to \Gamma$: $(x,\varphi) \longmapsto \varphi$ n'est pas surjective, ce qui démontre l'existence d'un morphisme φ injectif.

Démonstration de la proposition 6.2. Désignons par $\mathcal{M}$ l'ouvert des morphismes injectifs φ ci-dessus. Pour $\varphi \in \mathcal{M}$, le conoyau $E = E_\varphi$ est un fibré de rang 2 de classe de Chern, en posant $\ell = c_1 + 2 - \chi$, $u = c_1(\mathcal{O}(1))$

$$c(E) = \frac{1}{(1+u)^{\ell}(1-u)^{c_1+\ell}}$$

$$= 1 + c_1 u + c_2 u^2$$

Pour ce fibré, on a évidemment $H^1(\mathbb{P}_2, E) = 0$ et $H^0(\mathbb{P}_2, E(-1)) = 0$. Puisque $c_1 > 0$, ce fibré est stable, d'où la proposition 6.2.

Proposition 6.5.: Il existe dans $M(c_1, c_2)$ un ouvert de Zariski non vide correspondant à des fibrés engendrés par leurs sections globales si et seulement si l'une des conditions suivantes est réalisée :

$$c_1 > 0 \ \underline{\text{et}} \ \chi(c_1,c_2) \geq 4$$
$$(c_1,c_2) = (1,1) \ \underline{\text{ou}} \ (2,4)$$

Démonstration. Remarquons tout d'abord que si E est un fibré stable de rang 2 de classe de Chern $c_1 \leq 0$, il existe des droites $\ell \subset \mathbb{P}_2$ telles que $E|_\ell$ soit isomorphe à $\mathcal{O}(m) \oplus \mathcal{O}(c_1 - m)$, avec $m < 0$; par conséquent E n'est pas engendré par ses sections globales.

Supposons $c_1 > 0$. On se ramène par tensorisation par $\mathcal{O}(i)$ au cas où $1 \leq \chi \leq c_1 + 2$; il suffit dans ce cas de vérifier les assertions suivantes:

(1) si $4 \leq \chi \leq c_1+2$, ou si $(c_1,c_2) = (1,1)$ ou $(2,4)$, l'ouvert de W correspondant à des fibrés engendrés par leurs sections globales est non vide.

(2) si $1 \leq \chi \leq 3$, aucun fibré de W n'est engendré par ses sections globales, sauf si $(c_1,c_2) = (1,1)$ ou $(2,4)$. L'ouvert de W des fibrés E tels que E(1) soit endré par ses sections globales est non vide.

Démontrons **(1')** . On peut trouver un morphisme $\varphi \in \mathcal{M}$

$$\varphi = (\varphi_1, \varphi_2) : \mathcal{O}(-1)^{2c_1+2-\chi} \longrightarrow (Q^*)^{c_1+2-\chi} \oplus \mathcal{O}^{\chi}$$

tel que $\varphi_1 : \mathcal{O}(-1)^{2c_1+2-\chi} \longrightarrow (Q^*)^{c_1+2-\chi}$ soit surjectif ; ceci est possible en répétant l'argument utilisé dans la démonstration du lemme **6.4**, puisque le fibré $\underline{\mathrm{Hom}}(\mathcal{O}(-1)^{2c_1+2-\chi}, (Q^*)^{c_1+2-\chi})$ est engendré par ses sections globales, et que la différence des rangs est $2c_1 + 2 - \chi - 2(c_1+2-\chi) = \chi - 2 \geq 2$. Pour le fibré conoyau $E = E_\varphi$, le morphisme ev: $\mathcal{O}^{\chi} \longrightarrow E$ est alors surjectif.

Dans le cas $(c_1,c_2) = (1,1)$, on sait que $M(1,1)$ est réduit à un point, défini par le fibré Q, qui est engendré par ses sections; dans le cas $(c_1,c_2) = (2,4)$ il résulte du lemme 4.2 qu'un fibré E de classes de Chern $(2,4)$ générique a une présentation de la forme

$$0 \longrightarrow \mathcal{O}(-2) \longrightarrow \mathcal{O}^3 \longrightarrow E \longrightarrow 0$$

et par conséquent , un tel fibré est engendré par ses sections globales.

Démontrons **(2)** : le morphisme $ev: \mathcal{O}^\chi \longrightarrow E$ n'est évidemment pas surjectif pour $\chi = 1$, ni pour $\chi = 2$, car alors E serait trivial, ce qui contredit la stabilité. Si ce morphisme était surjectif pour $\chi = 3$, son noyau Z serait un fibré de rang 1 de classe de Chern $(-c_1, c_1^2 - c_2)$, ce qui impose $c_2 = c_1^2$. Ceci n'est possible que dans les cas $(c_1,c_2) = (1,1)$ ou $(2,4)$. Pour $\varphi \in \mathcal{M}$, on a un épimorphisme

$$Q^{c_1+2-\chi} \oplus \mathcal{O}(1)^\chi \longrightarrow E(1)$$

E désignant le conoyau du morphisme φ ; $E(1)$ est alors engendré par ses sections, ce qui démontre l'assertion (2), et par suite la proposition.

Proposition 6.6.: <u>Supposons</u> $c_1 > 0$ <u>et</u> $8 \leq \chi \leq c_1 + 2$. <u>L'ouvert de Zariski de</u> W <u>correspondant à des fibrés très amples est non vide</u>.

Démonstration. Il s'agit de montrer que l'ouvert de Zariski de W des fibrés E pour lesquels le morphisme canonique

$$j_E : \mathcal{O}^\chi = \mathcal{O} \otimes H^0(\mathbb{P}_2, E) \longrightarrow J^1E$$

est surjectif est non vide. On a déjà vu que les fibrés correspondant à un ouvert de Zariski $W_2 \subset W$ s'obtiennent comme conoyau d'un morphisme injectif $\psi : Z \to \mathcal{O}^\chi$, où Z est lui-même le noyau d'un épimorphisme

$$\mathcal{O}(-1)^{2c_1+2-\chi} \longrightarrow (Q^*)^{c_1+2-\chi}$$

Il suffit donc de montrer que l'on peut choisir ψ de telle sorte que le morphisme

$$\omega(\psi) : Z \longrightarrow T^* \otimes E$$

associé à la suite exacte $0 \to Z \overset{\psi}{\to} \mathcal{O}^\chi \to E \to 0$ et donné par la seconde forme fondamentale [6] soit surjectif. L'hypothèse $\chi \geq 8$ signifie que $\mathrm{rang}(Z) \geq 6$.

Soit $x \in \mathbb{P}_2$; désignons par $\omega(\psi, x) : Z(x) \longrightarrow T^* \otimes E(x)$ l'application $\mathbb{C}$-linéaire induite au-dessus du point x par $\omega(\psi)$. Soient d'autre part Ω l'ouvert des morphismes $\psi: Z \to \mathcal{O}^\chi$ injectifs, $\Sigma \subset \Omega \times \mathbb{P}_2$ le sous-ensemble algébrique des couples (ψ, x) tels que $\mathrm{rang}\ \omega(\psi,x) < 4$, $\Sigma(x)$ la fibre de Σ au-dessus du point $x \in \mathbb{P}_2$.

Lemme 6.7.: Pour tout $x \in \mathbb{P}_2$, on a $\dim \Sigma(x) \leq \dim \Omega - 3$

Ceci implique $\dim \Sigma < \dim \Omega$; par suite, la projection $pr_1 : \Sigma \to \Omega$ n'est pas surjective, et donc il existe $\psi \in \Omega$ tel que le morphisme $\omega(\psi, x)$ soit surjectif en tout point $x \in \mathbb{P}_2$, d'où la proposition **6.6.**

Démonstration du lemme 6.7.: Soient $x \in \mathbb{P}_2$, et, au voisinage de x, ∇ une connexion sur le fibré $\underline{\mathrm{Hom}}(Z, \mathcal{O}^{\chi})$ associée à une connexion sur Z, de sorte que dans l'identification $T^* \otimes \underline{\mathrm{Hom}}(Z, \mathcal{O}^{\chi}) \simeq \underline{\mathrm{Hom}}(Z, T^* \otimes \mathcal{O}^{\chi})$ on peut écrire

$$\omega(\psi, x) = p(\nabla\psi(x))$$

où $p : \underline{\mathrm{Hom}}(Z, T^* \otimes \mathcal{O}^{\chi}) \longrightarrow \underline{\mathrm{Hom}}(Z, T^* \otimes E)$ est la projection induite par la projection $\mathcal{O}^{\chi} \to E$.

Soit **E** le fibré universel défini au-dessus de $\Omega \times \mathbb{P}_2$ par le conoyau du morphisme canonique $\Omega \times Z \longrightarrow \Omega \times \mathcal{O}^{\chi}$; si $\mathbf{E}|\Omega' \times \{x\} \xrightarrow{\simeq} \Omega' \times \mathbb{C}^2$ est une trivialisation de $\mathbf{E}|\Omega' \times \{x\}$ au-dessus de l'ouvert Ω', on obtient par composition un morphisme surjectif, dépendant algébriquement de $\psi \in \Omega'$:

$$\xi(\psi) : \underline{\mathrm{Hom}}_x(Z, T^* \otimes \mathcal{O}^{\chi}) \longrightarrow \underline{\mathrm{Hom}}_x(Z, T^* \otimes \mathbb{C}^2)$$

On peut en outre choisir la trivialisation de sorte que $\xi(\psi)$ ne dépende que de la la valeur de ψ au point x. Sur l'ouvert Ω', $\Sigma(x) \cap \Omega'$ est donné par les $\psi \in \Omega'$ tels que

$$\mathrm{rang}\ \xi(\psi)(\nabla\psi(x)) < 4$$

Or, l'application $\Omega' \longrightarrow \underline{\mathrm{Hom}}_x(Z, T^* \otimes \mathbb{C}^2) : \psi \longmapsto \xi(\psi)(\nabla\psi(x))$ est une submersion. En effet, en restriction au sous-espace de Ω' des ψ' de la forme $\psi' = \psi + \sigma$, où $\sigma : Z \to \mathcal{O}^{\chi}$ est un morphisme tel que $\sigma(x) = o$, et où ψ est fixé, elle a pour dérivée au point ψ l'application linéaire

$$\sigma \longmapsto \xi(\psi)(\nabla\sigma(x))$$

Le fibré Z étant un sous-fibré de $\mathcal{O}(-1)^{2c_1+2-\chi}$, le fibré $\underline{\mathrm{Hom}}(Z, \mathcal{O}^{\chi})$ est très ample. Ceci montre que l'application $\sigma \longmapsto \nabla\sigma(x)$ est surjective, et il en est de même de la dérivée ci-dessus. Ceci démontre le lemme 6.7.

Proposition 6.8.: <u>On suppose</u> $c_1 > o$ <u>et</u> $1 \leq \chi \leq \mathrm{Min}(7, c_1+2)$. <u>Alors</u>:

(1) <u>Aucun fibré de</u> W <u>n'est très ample</u>

(2) <u>L'ouvert de</u> W <u>correspondant aux fibrés</u> E <u>tels que</u> E(1) <u>soit très ample est non vide, sauf si</u> $\chi(1) = \chi + c_1 + 4 \leq 7$, <u>c'est-à-dire si</u> $(c_1, c_2) = (1,2), (1,3)$ <u>ou</u> $(2,6)$

(3) <u>Si</u> $\chi(1) \leq 7$, <u>pour tout fibré</u> E <u>correspondant à un point de</u> W, E(2) <u>est très ample</u>.

Démonstration. Soit E un fibré correspondant à un point de W. Dire que E est très ample signifie que le morphisme canonique

$$j_E : \mathcal{O}^\chi = \mathcal{O} \otimes H^\circ(\mathbb{P}_2, E) \longrightarrow J^1E$$

est surjectif. Puisque J^1E est de rang 6, ce n'est pas possible si $1 \leqslant \chi < 6$. Pour $\chi = 6$, on aurait un isomorphisme de fibrés $\mathcal{O}^\chi \simeq J^1E$. Or, d'après le **3.5** on a, avec $u = c_1(\mathcal{O}(1))$

$$c(J^1E) = (1 + (c_1 - 2)\, u + (c_2 - c_1 + 1)\, u^2)^3$$

Si J^1E est trivial, on a obligatoirement $c_1 = 2$ et $c_2 = 1$, ce qui contredit la condition $c_2 - \frac{c^2}{4} \geqslant 2$ imposée par la stabilité. Pour $\chi = 7$, le noyau de j_E serait un fibré de rang 1, et $\frac{1}{c(E(-1))^3}$ serait la classe de Chern d'un fibré de rang 1. Or:

$$\frac{1}{c(E(-1))^3} = 1 - 3c_1(E(-1))\, u + 3\left[2c_1(E(-1))^2 - c_2(E(-1))\right] u^2$$

$$= 1 - 3(c_1 - 2)\, u + 3\,(2c_1^2 - 7c_1 + 7 - c_2) u^2$$

On aurait donc $2c^2 - 7c_1 + 7 = c_2$; du fait $\chi = 7$, ceci impose $c_1 = 3$ et $c_2 = 4$, ce qui est contraire à l'hypothèse $\chi \leqslant c_1 + 2$. Ceci démontre (1).

(2) Il s'agit de voir que l'ouvert de W des fibrés pour lesquels le morphisme canonique

$$j_{E(1)} = \mathcal{O}^{\chi(1)} \longrightarrow J^1E(1)$$

est surjectif est non vide, sauf si $\chi(1) \leqslant 7$. Si $\chi \geqslant 4$, on sait d'après la proposition **6.5** qu'il existe dans W des fibrés E engendrés par leurs sections; pour un tel fibré E, E(1) est très ample. Il reste à étudier les cas $\chi = 3, 2, 1$, qui seront l'objet des propositions **6.9**, **6.12** et **6.14**.

Proposition 6.9. (Cas $\chi = 3$) : <u>Si</u> $E = E_\varphi$ <u>est le conoyau d'un morphisme injectif</u>

$$\varphi = (\varphi_1, \varphi_2) : \mathcal{O}(-1)^{2c_1 - 1} \longrightarrow (Q^*)^{c_1 - 1} \oplus \mathcal{O}^3$$

<u>alors</u> E(1) <u>est très ample pour</u> φ <u>générique</u>.

Lemme 6.10.: <u>Soit</u> Ω_1 <u>l'ouvert de Zariski des sections</u> φ_1 <u>du fibré</u> $\mathrm{Hom}\left(\mathcal{O}^{2c_1 - 1}, Q^{c_1 - 1}\right)$ <u>vérifiant les conditions suivantes</u>

1°) <u>pour tout</u> $x \in \mathbb{P}_2$, $\varphi_1(x)$ <u>est de rang</u> $\geqslant 2c_1 - 3$

2°) <u>le morphisme</u> φ_1 <u>est transverse à la sous-variété de</u> $\mathrm{Hom}\left(\mathcal{O}^{2c_1 - 1}, Q^{c_1 - 1}\right)$ <u>des morphismes de rang</u> $2c_1 - 3$.

<u>Alors</u> Ω_1 <u>est un ouvert non vide</u>.

Démonstration. La difficulté est due au fait que le fibré Q n'est pas très ample. Pour $c_1 = 2$, le résultat est évident, car il existe un morphisme surjectif $\mathcal{O}^3 \longrightarrow Q$.

On supposera donc $c_1 > 2$. Réaliser la condition 1°) se fait par un argument analogue à celui du lemme **6.4.** Soit Ω_2 l'ouvert de Zariski des morphismes φ_1 satisfaisant à cette condition, et soit Y le sous-espace de $\Omega_2 \times \mathbb{P}_2$ des couples (φ_1, x) tels que

$$\text{rang } \varphi_1(x) = 2c_1 - 3$$

C'est une variété lisse et connexe de dimension $N = \dim \Omega_2$. L'existence d'un morphisme $\varphi_1 \in \Omega_2$ vérifiant la condition 2°) est alors une conséquence immédiate du fait suivant: le sous-espace $Y' \subset Y$ des couples (φ_1, x) tels que φ_1 ne soit pas transverse en x à la sous-variété des morphismes de rang $2c_1 - 3$ est de codimension 1. Pour le voir, il suffit de vérifier que $Y' \neq Y$.

Supposons d'abord $c_1 = 3$. L'espace vectoriel dont est issu $\mathbb{P}_2$ s'identifie à l'espace vectoriel $V = \Gamma(\mathbb{P}_2, Q)$ des sections de Q; si (e_1, e_2, e_3) est une base de V, chacune des matrices A_i :

$$A_1 = \begin{pmatrix} 1 & 0 & 0 & 0 & 0 \\ 0 & 1 & 0 & 0 & 0 \\ 0 & 0 & 0 & 0 & 1 \end{pmatrix}.$$

$$A_2 = \begin{pmatrix} 0 & 0 & 1 & 0 & 0 \\ 1 & 0 & 0 & 0 & 0 \\ 0 & 0 & 0 & 1 & 0 \end{pmatrix}$$

définit un morphisme $\psi_i : \mathcal{O}^5 \longrightarrow Q$. Si on rapporte Q au voisinage de $x_o = (o,o,1)$ au repère local associé à (e_1, e_2), la matrice du morphisme $\psi : \mathcal{O}^5 \longrightarrow Q^2$ s'écrit audessus du point $z = (z_1 e_1 + z_2 e_2 + e_3)$

$$\psi(z) = \begin{pmatrix} 1 & 0 & 0 & 0 & -z_1 \\ 0 & 1 & 0 & 0 & -z_2 \\ 0 & 0 & 1 & -z_1 & 0 \\ 1 & 0 & 0 & -z_2 & 0 \end{pmatrix}$$

Au-dessus du point $x_o = (o,o,1)$, $\psi(x_o) = \psi_o$ est de rang 3; l'espace tangent en ψ_o à la variété des applications linéaires $h = (h_i^j) : \mathbb{C}^5 \to \mathbb{C}^4 = Q^2_{x_o}$ de rang 3 est donnée par les équations

$$h_4^4 - h_4^1 = o \quad ; \quad h_5^4 - h_5^1 = o$$

Puisque $\left((\psi(z)_4^4 - \psi(z)_4^1, \psi(z)_5^4 - \psi(z)_5^1\right) = (-z_2, z_1)$, on voit que la condition de transversalité est bien satisfaite pour ψ au point x_o, et par suite $(\psi, x_o) \in Y - Y'$.

Passons au cas $c_1 > 3$. Soit $\epsilon : \mathcal{O}^2 \longrightarrow Q$ le morphisme défini par les sections associées à e_1 et e_2. On écrit

$$\mathcal{O}^{2c_1-1} = \mathcal{O}^5 \oplus \underbrace{\mathcal{O}^2 \oplus \ldots \oplus \mathcal{O}^2}_{c_1-3}$$

$$Q^{c_1-1} = Q^2 \oplus \underbrace{Q \oplus \ldots \oplus Q}_{c_1-3}$$

Le morphisme $\varphi_1 : \mathcal{O}^{2c_1-1} \longrightarrow Q^{c_1-1}$ de matrice

$$\begin{pmatrix} \psi & 0 & \cdots & 0 \\ 0 & e & & \vdots \\ \vdots & & \ddots & 0 \\ 0 & \cdots & 0 & \epsilon \end{pmatrix}$$

satisfait à la condition de transversalité au point x_o. D'où le lemme **6.10.**

Démonstration de la proposition 6.9.: Soit $\varphi = (\varphi_1, \varphi_2)$ un morphisme injectif $\mathcal{O}(-1)^{2c_1-1} \longrightarrow (Q^*)^{c_1-1} \oplus \mathcal{O}^3$ tel que φ_1 satisfasse aux conditions du lemme **6.10.** Le sous-espace $X \subset \mathbb{P}_2$ défini par l'annulation de $\Lambda^{2c_1-2}\varphi_1(x)$:

$$X = \{x \in \mathbb{P}_2 \text{ , rang } \varphi_1(x) = 2c_1-3\}$$

est réduit à un nombre fini de points simples. Ceci entraîne en particulier que le $\mathcal{O}$-module $\mathcal{G}$ conoyau du morphisme φ_1 est de support fini. On peut dire plus:

Lemme 6.11.: <u>Le faisceau $\mathcal{G}$ est un $\mathcal{O}_X$-module libre de rang 1.</u>

Démonstration. Soient $x \in X$, et m_x l'idéal maximal de $\mathcal{O}_x$. Vu la condition de transversalité, on peut choisir des repères locaux dans $\mathcal{O}(-1)^{2c_1-1}$ et $(Q^*)^{c_1-1}$ au voisinage de x de telle sorte que la matrice de φ_1 s'écrive au voisinage de x

$$\begin{pmatrix} 1 & 0 & \cdots & 0 & 0 & 0 \\ 0 & 1 & & & \vdots & \vdots \\ \vdots & & \ddots & & \vdots & \vdots \\ 0 & & & 1 & 0 & 0 \\ 0 & \cdots & \cdots & 0 & \xi & \eta \end{pmatrix}$$

où (ξ, η) induisent une base dans $T^*_x = m_x / m_x^2$ et donc donnent un système de générateurs de m_x. Il en résulte que le $\mathcal{O}_x$-module $\mathcal{G}_x$ est isomorphe à $\mathcal{O}_{x/m_x}$, ce qui démontre le lemme **6.11.**

Fin de la démonstration de la proposition 6.9. On a pour le fibré $E = E_\varphi$ la suite exacte $\mathcal{O}^3 \xrightarrow{ev} E \longrightarrow \mathcal{G} \longrightarrow o$. Considérons le diagramme commutatif

$$\begin{array}{ccccc} H^o(\mathbb{P}_2, \mathcal{O}(1)^3) & \longrightarrow & H^o(\mathbb{P}_2, E(1)) & & \\ \downarrow j_1 = j_{\mathcal{O}(1)^3} & & \downarrow j = j_{E(1)} & & \\ J^1\mathcal{O}(1)^3(x) & \longrightarrow & J^1E(1)(x) & \longrightarrow J^1\mathcal{G}(1)(x) & \longrightarrow o \end{array}$$

dans lequel la deuxième ligne est exacte. Puisque $\mathcal{O}(1)$ est très ample, la flèche j_1 est surjective pour tout $x \in \mathbb{P}_2$; en dehors de $X = \mathrm{supp}(\mathcal{G})$, la flèche j est donc surjective. Pour $x \in X$, on a $J^1\mathcal{G}(1)(x) \simeq \mathcal{G}(1)(x) \simeq \mathbb{C}$; pour voir que j est surjective, il suffit donc de vérifier qu'il existe une section s de $E(1)$ telle que l'image du jet $j(s)(x)$ soit non nulle dans $J^1\mathcal{G}(1)(x)$. En vertu de la suite exacte

$$\mathcal{O}(1)^3(x) \longrightarrow E(1)(x) \longrightarrow \mathcal{G}(1)(x) \longrightarrow o$$

il revient au même de demander que la valeur $s(x)$ de s au point x ne soit pas dans l'image de l'application linéaire $\mathcal{O}(1)^3(x) \longrightarrow E(1)(x)$. Une telle section existe car $E(1)$ est clairement engendré par ses sections globales. Donc $E(1)$ est très ample.

Proposition 6.12. (Cas $\chi=2$) : Soit $E = E_\varphi$ le conoyau d'un morphisme injectif

$$\varphi = (\varphi_1, \varphi_2) : \mathcal{O}(-1)^{2c_1} \longrightarrow (Q^*)^{c_1} \oplus \mathcal{O}^2$$

Alors : (1) $E(1)$ est très ample pour φ générique, sauf si $c_1 = 1$.

(2) Si $c_1 = 1$, $E(2)$ est très ample pour tout φ.

Démonstration. On supposera ici encore que la section du fibré $\mathrm{Hom}\left(\mathcal{O}(-1)^{2c_1}, (Q^*)^{c_1}\right)$ définie par φ_1 est transverse à la sous-variété des morphismes de rang $2c_1-1$, et ne coupe pas le sous-espace des morphismes de rang $< 2c_1-1$. Dans ces conditions, la courbe X d'équation $\det \varphi_1 = o$ est une courbe lisse de degré c_1, et le conoyau $\mathcal{G}$ de φ_1 est un $\mathcal{O}_X$-module localement libre de rang 1: on a donc

$$\mathcal{G} = i_* \mathcal{O}_X(L)$$

où L est un fibré vectoriel de rang 1 sur X, et $i : X \hookrightarrow \mathbb{P}_2$ le morphisme d'inclusion. D'autre part, on a la suite exacte

$$0 \longrightarrow \mathcal{O}^2 \xrightarrow{ev} E \longrightarrow \mathcal{G} \longrightarrow 0$$

Lemme 6.13. Sur la courbe X, le fibré $L' = L(1)$ est très ample si $c_1 > 1$,

Démonstration. Soit $x \in X$. En vertu de la suite exacte

$$H^{o}(X,L') \longrightarrow J^{1}L'(x) \longrightarrow H^{1}(X,L'(-2x))$$

il suffit de prouver que $H^1(X,L'(-2x)) = o$. La tangente à X en x recoupe X en $y_3,\dots,y_{c_1}$, et on a $\mathcal{O}(1)|_X \simeq \mathcal{O}_X\left(2x + \sum_{3\leqslant i} y_i\right)$. Par conséquent, $L'(-2x)=L\left(\sum_{3\leqslant i} y_i\right)$.

Par suite, $\dim H^1(X,L'(-2x)) \leqslant \dim H^1(X,L)$. Or, $H^1(X,L)=H^1(\mathbb{P}_2,\mathcal{G})=o$ d'après la suite exacte définissant $\mathcal{G}$. Ceci démontre le lemme **6.13**.

Fin de la démonstration de la proposition 6.12. Compte tenu de la suite exacte

$$0 \longrightarrow \mathcal{O}(1)^2 \longrightarrow E(1) \longrightarrow \mathcal{G}(1) \longrightarrow 0$$

il suffit de recopier l'argument de la proposition **5.5**: on en déduit que E(1) est très ample.

Le cas $c_1=1$ échappe au raisonnement ci-dessus. On sait d'ailleurs que dans ce cas, $c_1(E(1))=3$, $c_2(E(1))=4$, et d'après le corollaire **3.3**, E(1) n'est jamais ample. Cependant, pour tout fibré de M(1,2), on a d'après la proposition **3.4** $H^1(\mathbb{P}_2,E)=o$, d'où il résulte que E(1) est engendré par ses sections; par suite E(2) est très ample.

Proposition 6.14. (Cas $\chi=1$) : On suppose $c_1 > o$ et $\chi=1$

(1) Tout fibré E de W est conoyau d'un morphisme injectif

$$\psi : \mathcal{O}(-2)^{c_1+1} \longrightarrow \mathcal{O} \oplus \mathcal{O}(-1)^{c_1+2}$$

(2) Pour ψ générique, E(1) est très ample, sauf dans les cas $(c_1,c_2)=(1,3)$ et $(c_1,c_2)=(2,6)$

(3) Dans les cas $(c_1,c_2)=(1,3)$ ou $(2,6)$, E(2) est très ample pour tout ψ.

Démonstration. (1) On a $\chi(1)=c_1+5$. La suite spectrale de Beilinson appliquée à E(1) montre que pour tout fibré E correspondant à un point de W on a les suites exactes de $\mathcal{O}$-modules

$$0 \longrightarrow Q^* \xrightarrow{d_1^{-1,0}} \mathcal{O}^{c_1+5} \longrightarrow F \longrightarrow 0$$

$$0 \longrightarrow \mathcal{O}(-1)^{c_1+1} \xrightarrow{d_2^{-2,1}} F \longrightarrow E(1) \longrightarrow 0$$

Il résulte de la seconde suite exacte que F est $\mathcal{O}$-module localement libre de rang c_1+3. En fait $F \simeq \mathcal{O}(1) \oplus \mathcal{O}^{c_1+2}$. En effet, le morphisme $d_1^{-1,0}$ est injectif comme morphisme de fibrés; il est donné par une application linéaire injective : $V^* \longrightarrow \mathbb{C}^{c_1+5}$. Il en résulte qu'il existe $g \in GL(c_1+5,\mathbb{C})$ rendant commutatif le diagramme

$$
\begin{array}{ccc}
 & & {}^{c_1+5} \\
 & \overset{d_1^{-1,0}}{\nearrow} & \\
Q^* & & \big\downarrow g \\
 & \underset{\tau}{\searrow} & \\
 & & {}^{c_1+5}
\end{array}
$$

où τ est le morphisme $Q^* \xrightarrow{(i,o)} \mathcal{O}^{c_1+5} = \mathcal{O}^3 \oplus \mathcal{O}^{c_1+2}$ associé à l'inclusion canonique $i : Q^* \hookrightarrow \mathcal{O}^3$. Les conoyaux de $d_1^{-1,0}$ et τ sont isomorphes, ce qui donne l'assertion ci-dessus; ceci démontre (1).

(2) Réciproquement, si ψ est un morphisme injectif de fibrés

$$\psi = (\psi_1, \psi_2) : \ \mathcal{O}(-2)^{c_1+1} \longrightarrow \mathcal{O} \oplus \mathcal{O}(-1)^{c_1+2}$$

son conoyau E est un fibré stable de rang 2 de classes de Chern c_1 et c_2 . On peut ici encore choisir ψ_2 de telle sorte que la section du fibré $\underline{\mathrm{Hom}}\left(\mathcal{O}(-2)^{c_1+2}, \mathcal{O}(-1)^{c_1+2}\right)$ qu'il définit soit partout de rang $\geq c_1$ et rencontre transversalement la variété des morphismes de rang c_1 . Soit Ω l'ouvert des morphismes injectifs $\psi = (\psi_1, \psi_2)$ qui satisfont à cette condition de transversalité.

Le sous-espace X de $\Omega \times \mathbb{P}_2$ défini par l'annulation du morphisme

$$(\psi, x) \longmapsto \Lambda^{c_1+1} \psi_2(x) :$$

$$X = \{(\psi, x) \in \Omega \times \mathbb{P}_2 \ , \ \mathrm{rang}\ \psi_2(x) = c_1\}$$

est un revêtement fini non ramifié de Ω ; c'est donc une variété lisse. Soit $\mathcal{G}$ le $\mathcal{O}$-module conoyau de $\psi_2(1)$; on a le diagramme commutatif

$$
\begin{array}{ccccccccc}
0 & \longrightarrow & \mathcal{O}(-1)^{c_1+1} & \xrightarrow{\psi_2(1)} & \mathcal{O}^{c_1+2} & \longrightarrow & \mathcal{G} & \longrightarrow & 0 \\
 & & \big\downarrow {\scriptstyle -\psi_1(1)} & & \big\downarrow & & \| & & \\
0 & \longrightarrow & \mathcal{O}(1) & \longrightarrow & E(1) & \longrightarrow & \mathcal{G} & \longrightarrow & 0
\end{array}
$$

dans lequel les lignes sont exactes. Le $\mathcal{O}$-module $\mathcal{G}$ n'est pas localement libre; on a

$$\mathrm{rang}\ \mathcal{G}(x) = \begin{cases} 1 & \text{si } (\psi, x) \notin X \\ 2 & \text{si } (\psi, x) \in X \end{cases}$$

Bien que $\mathcal{G}$ ne soit pas localement libre, la première suite exacte permet encore de définir un morphisme "deuxième forme fondamentale" pour tout $x \in \mathbb{P}_2$:

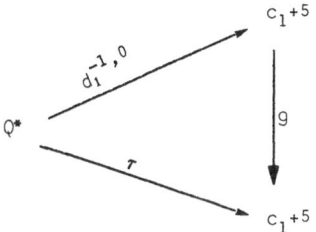

$$\omega(\psi,x) : \mathcal{O}(-1)^{c_1+1}(x) \longrightarrow T^* \otimes \mathcal{G}(x)$$

rendant commutatif le diagramme

$$\begin{array}{ccccccc}
\mathcal{O}(-1)^{c_1+1}(x) & \longrightarrow & \mathbb{C}^{c_1+2} = \mathcal{O}^{c_1+2}(x) & \longrightarrow & \mathcal{G}(x) & \longrightarrow & 0 \\
\downarrow{\scriptstyle \omega(\psi,x)} & & \downarrow & & \| & & \\
T^* \otimes \mathcal{G}(x) & \longrightarrow & J^1 \mathcal{G}(x) & \longrightarrow & \mathcal{G}(x) & \longrightarrow & 0
\end{array}$$

Posons

$$\Sigma_1 = \{(\psi,x) \in X,\ \operatorname{rang} \omega(\psi,x) < 4\}$$

$$\Sigma_2 = \{(\psi,x) \in \Omega \times \mathbb{P}_2 - X,\ \operatorname{rang} \omega(\psi,x) < 2\}$$

Du fait que le fibré $\underline{\mathrm{Hom}}\left(\mathcal{O}(-1)^{c_1+1}, \mathcal{O}^{c_1+2}\right)$ est très ample, on déduit exactement comme dans le lemme **6.7** que $\operatorname{codim}_X \Sigma_1 \geqslant c_1-2$ et $\operatorname{codim}_\Omega \Sigma_2(x) \geqslant c_1$, où $\Sigma_2(x)$ est la fibre de Σ_2 au-dessus du point $x \in \mathbb{P}_2$; par suite, dans l'ouvert $U = \Omega \times \mathbb{P}_2 - X$, on a $\operatorname{codim} \Sigma_2 \geqslant c_1-2$. Si $c_1 \geqslant 3$, il en résulte que la projection $(\psi,x) \longmapsto \psi : \Sigma_1 \cup \Sigma_2 \longrightarrow \Omega$ n'est pas surjective; si on prend ψ dans le complémentaire de l'image, $\omega(\psi,x)$ sera surjective pour tout $x \in \mathbb{P}_2$, et il en sera de même du morphisme canonique

$$j_{\mathcal{G}} : \mathbb{C}^{c_1+2} = H^\circ(\mathbb{P}_2, \mathcal{G}) \longrightarrow J^1\mathcal{G}(x)$$

A la suite exacte $0 \to \mathcal{O}(1) \to E(1) \to \mathcal{G} \to 0$ est associé un diagramme commutatif pour tout $x \in \mathbb{P}_2$

$$\begin{array}{ccccccc}
H^\circ(\mathbb{P}_2, \mathcal{O}(1)) & \longrightarrow & H^\circ(\mathbb{P}_2, E(1)) & \longrightarrow & H^\circ(\mathbb{P}_2, \mathcal{G}) & \longrightarrow & 0 \\
\downarrow{\scriptstyle j_{\mathcal{O}(1)}} & & \downarrow{\scriptstyle j_{E(1)}} & & \downarrow{\scriptstyle j_{\mathcal{G}}} & & \\
J^1\mathcal{O}(1)(x) & \longrightarrow & J^1E(1)(x) & \longrightarrow & J^1\mathcal{G}(x) & \longrightarrow & 0
\end{array}$$

dans lequel les flèches $j_{\mathcal{O}(1)}$ et $j_{\mathcal{G}}$ sont surjectives: par suite $j_{E(1)}$ est surjective, ce qui démontre que $E(1)$ est très ample.

Cas particuliers. Si $c_1 = 1$, alors $c_2=3$; $c_1(E(1))=3$ et $\chi(E(1))=6$; le fibré $E(1)$ n'est jamais ample d'après le corollaire **3.3.** Si $c_1=2$, $c_2=6$; alors $c_1(E(1))=4$ et $\chi(E(1))=7$; par suite, $E(1)$ n'est jamais très ample d'après la proposition **3.6.**

Cependant, pour tout ψ, $E(1)$ est engendré par ses sections, et par suite $E(2)$ est très ample. Ceci démontre la proposition **6.14.**

Démonstration du théorème 1.3.: Il suffit de choisir k tel que $c_1 + 2k > o$ et $1 \leqslant \chi(k) \leqslant c_1+2k+2$, et d'appliquer à $M(c_1+2k, c_2+kc_1+k^2)$ les propositions **6.6** et **6.8**.

Remarque 6.15. La condition $\chi \geqslant 8$ n'est nullement nécessaire pour qu'un fibré stable de rang 2 de classe de Chern c_1 et c_2 soit très ample. Par exemple, si p est entier $> o$, le conoyau d'un morphisme injectif

$$\mathcal{O}(-p+1) \longrightarrow \mathcal{O}(1)^3$$

est un fibré de rang 2 stable et très ample; pour ce fibré, on a

$$\chi = 9 - \tfrac{1}{2}(p-2)(p-3)$$

et par suite $\chi < 8$ dès que $p \geqslant 5$.

7. UN CRITERE D'AMPLITUDE POUR LES FIBRES STABLES GÉNÉRIQUES

Lemme 7.1.: Soit E un fibré vectoriel sur $\mathbb{P}_2$. Alors E est ample si et seulement si il existe un entier n tel que

$$H^1(\mathbb{P}_2, S^n E(-2)) = o \quad \underline{\text{et}} \quad H^2(\mathbb{P}_2, S^n E(-3)) = o$$

Démonstration. D'après [6], si E est ample, ces conditions sont réalisées pour n assez grand. Réciproquement, supposons ces conditions vraies pour l'entier n, et vérifions qu'alors $S^n E$ est très ample. Soit $m_x \subset \mathcal{O}$ l'idéal définissant le point x. On a alors la résolution

$$0 \longrightarrow \mathcal{O}(-3)^2 \longrightarrow \mathcal{O}(-2)^3 \longrightarrow m_x^2 \longrightarrow 0$$

La suite exacte de cohomologie associée montre que $H^1(\mathbb{P}_2, S^n E \otimes m_x^2) = o$, ce qui démontre que $S^n E$ est très ample.

Proposition 7.2.: Soit E un fibré stable de rang 2 sur $\mathbb{P}_2$, de classes de Chern c_1 et c_2. Alors E est ample si et seulement si

(1) $c_1 > 2$,

(2) il existe un entier n tel que $H^1(\mathbb{P}_2, S^n E(-2)) = o$

Démonstration. On a déjà vu que ces conditions sont nécessaires. Pour voir qu'elles sont suffisantes, il suffit de vérifier que $H^2(\mathbb{P}_2, S^n E(-3)) = o$ pour tout entier n.

Si $\ell \subset \mathbb{P}_2$ est une droite générique de $\mathbb{P}_2$, on a d'après le théorème de Grauert-Mülich $E|_\ell \simeq \mathcal{O}(m') \oplus \mathcal{O}(m'')$ avec m' et m'' entiers > 0. Par suite $H^1(\ell, S^n E(i)) = 0$ pour $i \geqslant -2$. La suite exacte de cohomologie associée à ℓ montre que tout entier n et $i \geqslant -2$

$$H^2(\mathbb{P}_2, S^n E(i-1)) \simeq H^2(\mathbb{P}_2, S^n E(i))$$

Puisque ce groupe s'annule pour i assez grand d'après le théorème B de J.P.Serre, on en déduit que $H^2(\mathbb{P}_2, S^n E(-3)) = 0$.

Corollaire 7.3.: Dans $M(c_1,c_2)$, les points correspondant à un fibré ample forment un ouvert de Zariski.

Démonstration. Ceci résulte du théorème de semi-continuité.

Corollaire 7.4. : Soit p un entier $\geqslant 7$. Alors il existe dans $M(p+2,p^2+p+1)$ des fibrés amples qui n'ont pas de sections globales non nulles.

Ce phénomène n'est pas étonnant; il est déjà facile de l'observer sur des courbes.

Démonstration. Le conoyau d'un morphisme injectif $\mathcal{O}(-p+1) \longrightarrow \mathcal{O}(1)^3$ est très ample, donc ample. Par suite, dans $M(p+2,p^2+p+1)$ les points représentant un fibré ample forment d'après le corollaire **7.3** un ouvert de Zariski non vide. Or, d'après la proposition **6.1**, si

$$\chi = 9 - \tfrac{1}{2}(p-2)(p-3) < 0$$

l'ouvert de Zariski correspondant à des fibrés E tels que $H^0(\mathbb{P}_2,E)=0$ est non vide. Ces deux ouverts de Zariski se rencontrent puisque $M(c_1,c_2)$ est irréductible, d'où le corollaire **7.4**.

On se propose maintenant de montrer que si $c_1 \geqslant 4$ et $\chi(c_1,c_2) \geqslant 4$, les fibrés génériques de $M(c_1,c_2)$ sont amples. Ceci repose sur la proposition suivante, due à Gieseker, qui donne un critère d'amplitude pour les fibrés engendrés par leurs sections globales [5].

Proposition 7.5.: Soit E un fibré vectoriel engendré par ses sections globales au-dessus d'une variété projective lisse X. Alors les conditions suivantes sont équivalentes:

(1) E est ample

(2) Pour toute courbe réduite $C \subset X$ et tout fibré quotient L de rang un de $E|_C$, L est non trivial

(3) Pour toute courbe réduite $C \subset X$, on a $H^o(C,E^*)=o$

Proposition 7.6.: On suppose $c_1 \geqslant 4$ et $\chi \geqslant 4$. Alors l'ouvert de Zariski de $M(c_1,c_2)$ des points correspondant à des fibrés amples est non vide.

Démonstration. **a)** Supposons d'abord $c_1 \geqslant 4$ et $4 \leqslant \chi \leqslant c_1+2$. On sait qu'il existe des épimorphismes

$$\mathcal{O}(-1)^{2c_1+2-\chi} \longrightarrow (Q^*)^{c_1+2-\chi}$$

Soit Z le noyau d'un tel épimorphisme; comme dans la proposition **6.6**, il existe des morphismes injectifs ψ: $Z \to \mathcal{O}^\chi$. Pour que le conoyau E d'un tel morphisme ψ soit ample, il suffit d'après la proposition **7.5** de choisir ψ de telle sorte que pour toute courbe réduite $C \subset \mathbb{P}_2$, on ait $H^o(C,E^*)=o$. Il revient au même de choisir ψ de telle sorte que pour tout $v \in \mathbb{C}^\chi = H^o(\mathbb{P}_2, \mathcal{O}^\chi)$, $v \neq o$, la section ${}^t\psi\cdot v$, image de v par le morphisme transposé ${}^t\psi$: $\mathcal{O}^\chi \longrightarrow Z^*$ ne s'annule sur aucune courbe.

Or, la suite exacte

$$0 \longrightarrow Q^{c_1+2-\chi} \longrightarrow \mathcal{O}(1)^{2c_1+2-\chi} \longrightarrow Z^* \longrightarrow 0$$

montre que pour $i \leqslant o$, l'espace des sections $\Gamma(\mathbb{P}_2, Z^*(i))$ a pour dimension

i	-3	-2	-1	0
$\dim \Gamma(\mathbb{P}_2, Z^*(i))$	0	$c_1+2-\chi$	$2c_1+2-\chi$	$3c_1$

Considérons pour $i=-2$ et -1 les morphismes

$$P(\Gamma(\mathbb{P}_2,\mathcal{O}(-i))) \times P(\Gamma(\mathbb{P}_2,Z^*(i))) \longrightarrow P(\Gamma(\mathbb{P}_2,Z^*))$$

induits par la multiplication $(f,s) \longmapsto fs$. Ces morphismes sont propres, et leur image Σ_i est un sous-ensemble algébrique fermé dont on peut minorer la codimension, suivant les valeurs de i, par

$$3c_1-1-(2c_1-\chi+3) = c_1-4+\chi \geqslant \chi \qquad \text{si } i = -1$$
$$3c_1-1-(c_1+6-\chi) = 2c_1-7+\chi \geqslant \chi \qquad \text{si } i = -2$$

Il en résulte que pour ψ générique, l'image du morphisme induit par ${}^t\psi$: $P_{\chi-1}(\mathbb{C}) \longrightarrow P(\Gamma(\mathbb{P}_2,Z^*))$ ne rencontre pas les sous-ensembles Σ_i ci-dessus. Ceci signifie exactement que pour tout $v \in \mathbb{C}^\chi$, $v \neq o$, la section ${}^t\psi\cdot v$ de Z^* ne s'annule sur aucune courbe.

b) Supposons maintenant $\chi > c_1+2$. Ceci implique ou bien $\chi \geqslant 8$, auquel cas on a déjà démontré (théorème **1.3**) que, génériquement, E est très ample, ou bien, éventuellement $c_1=4$ et $\chi=7$. Dans ce cas, on est ramené à considérer les fibrés E conoyaux d'un morphisme injectif (proposition **6.14**)

$$\psi = (\psi_1,\psi_2) : \mathcal{O}(-1)^3 \longrightarrow \mathcal{O}(1) \oplus \mathcal{O}^4$$

Pour que $H^o(C,E^*) = o$ pour toute courbe réduite $C \subset \mathbb{P}_2$, il suffit de choisir ψ_2 de telle sorte que pour tout $v \in \mathbb{C}^4-\{o\}$, l'image ${}^t\psi_2 \cdot v$ de v par le morphisme

$${}^t\psi_2 : \mathbb{C}^4 = \Gamma(\mathbb{P}_2,\mathcal{O}^4) \longrightarrow \Gamma(\mathbb{P}_2,\mathcal{O}(1))^3$$

ne s'annule sur aucune courbe de $\mathbb{P}_2$. Considérons le morphisme propre induit par la multiplication

$$\mathbb{P}_3 \times P(\Gamma(\mathbb{P}_2,\mathcal{O}(1)) \longrightarrow P(\Gamma(\mathbb{P}_2,\mathcal{O}(1))^3$$

Son image Σ est de codimension $\geqslant 4$. Il en résulte que pour ψ générique, l'image du morphisme $\mathbb{P}_3 \longrightarrow P(\Gamma(\mathbb{P}_2,\mathcal{O}(1))^3$ induit ${}^t\psi_2$ ne rencontre pas l'ensemble Σ. Pour un tel ψ, et pour $v \in \mathbb{C}^4-\{o\}$, ${}^t\psi_2 \cdot v$ ne s'annule sur aucune courbe, et par suite, d'après la proposition **7.5**, E est ample. Ceci démontre la proposition **7.6**.

Bibliographie

[1] BARTH W.: Some properties of rank-2 bundles on $\mathbb{P}_n$. Math.Ann.226,125-150 (1977).

[2] BARTH W.: Moduli of vector bundles on the projective plane.Inventiones math. 42, 63-91 (1977).

[3] BEILINSON A.: Coherent sheaves on $\mathbb{P}_n$ and problems of linear algebra. J.Functional Analysis,12, 68-69 (1978).

[4] FULTON W.: Ample vector bundles, Chern classes and numerical criteria. Inventiones math. 32, 171-178 (1976).

[5] GIESEKER D.: p-ample bundles and their Chern classes. Nagoya Math.J.43,91-116 (1971).

[6] GRIFFITHS P.A.: Hermitian differential geometry, Chern classes and positive vector bundles. Global Analysis(Papers in honor of Kodaïra). Princeton U.Press (1970).

[7] HARTSHORNE R.: Ample vector bundles on curves. Nagoya Math.J.43, 73-89 (1971).

[8] KLEIMAN S.L.: Ample vector bundles on surfaces. Proc.Amer.Math.Soc.3,673-676 (1969).

[9] LE POTIER J.: Fibrés stables de rang 2 sur $\mathbb{P}_2(\mathbb{C})$. Math.Ann.241, 217-256 (1979).

[10] LE POTIER J.: Sur le groupe de Picard de l'espace des modules des fibrés stables de rang 2 sur $\mathbb{P}_2$. (en préparation).

[11] SCHNEIDER M.: Stabile Vektorraumbündel vom Rang 2 auf der projektiven Ebene. Nachrichten der Akademie der Wissenschaften in Göttingen,6,83-86 (1976).

[12] SCHWARZENBERGER R.L.E.: Vector bundles on the projective plane.Proc.London Math.Soc. 11, 623-640,(1961).

[13] HULEK K.: Stable rank-2 vector bundles on $\mathbb{P}_2$ with c_1 odd. Math.Ann.,242, 241-266,(1979).

Zur Berechnung von Yang-Mills Potentialen
durch
holomorphe Vektorbündel

von
Günther Trautmann

Diese Arbeit ist aus Vorträgen des Autors in einem gemeinsamen Seminar zwischen Physikern und Mathematikern an der Universität Kaiserslautern hervorgegangen. Ihr erster Teil besteht aus einem Überblick über die Differentialgeometrie der Yang-Mills Gleichungen sowie einem Beweis des Satzes von Atiyah-Ward, [4], über die Korrespondenz zwischen selbstdualen euklidischen SU(2)-Yang-Mills Feldern und gewissen holomorphen Vektorbündeln über $\mathbb{P}_3(\mathbb{C})$.
Im zweiten Teil wird die Darstellung der holomorphen Vektorbündel von [22] benutzt, um ein Verfahren zu expliziten Berechnung der Yang-Mills Potentiale und Felder zu gewinnen. Dabei gehen die Koeffizienten, die die Moduli der zugehörigen holomorphen Vektorbündel bilden, über in die Koeffizienten der rationalen Ausdrücke in den Variablen des $\mathbb{R}^4$, die die Potentiale beschreiben. In § 4 sind Ergebnisse von [22], [23] zusammengestellt. Die Parametrisierung $\mathfrak{M}_k \to M_I(0,k)$, 4.9, entspricht dabei der Konstruktion der Bündel durch Monaden, [7]; hier operiert aber lediglich noch die Gruppe GL(k,$\mathbb{C}$) auf $\mathfrak{M}_k$. In § 5 werden die Bedingungen der reellen Struktur für die Instantonenbündel in Bedingungen für die Modul-Koeffizienten transformiert.
In § 6 werden die Yang-Mills Potentiale zur Instantonenzahl 1 (Lösungen von t'Hooft) abgeleitet. In den Fällen höherer Instantonenzahl kann man die Lösungen analog ableiten, wobei aber kompliziertere Nebenbedingungen für die Koeffizienten zu berücksich-

tigen sind, vgl. (4.8.4), (iv). Diese Nebenbedingungen können in zufriedenstellender Weise aber erst dann eliminiert werden, wenn hinreichend gute Information über die Modulräume der holomorphen Vektorbündel vorliegen, vgl. 4.9, 4.10. Die Arbeit ist wie folgt eingeteilt:

I. Yang-Mills Felder und holomorphe Vektorbündel.

§ 1 Zusammenhänge und Yang-Mills Gleichungen.

In diesem Paragraphen wird zunächst an den Begriff eines linearen Zusammenhanges in einem Vektorbündel erinnert, sodann werden einige fundamentale Dinge der Chern-Weil Theorie zusammengestellt und die Yang-Mills Felder als Krümmungen von linearen Zusammenhängen in Vektorbündeln interpretiert. Differenzierbar soll stets als unendlich oft differenzierbar verstanden werden.

(1.1) Ist M eine differenzierbare Mannigfaltigkeit, so bezeichne T bzw. T^* das komplexifizierte Tangential- bzw. Cotangentialbündel. Ist E ein differenzierbares komplexes Vektorbündel auf M mit Faser $\mathbb{C}^r$, so ist ein (linearer) Zusammenhang in E ein $\mathbb{C}$-linearer Vektorbündel-Homomorphismus

$$\nabla : E \to T^* \otimes E$$

mit der Eigenschaft

$$\nabla(fs) = f\nabla(s) + df \otimes s \ ,$$

wobei s einen (lokalen) differenzierbaren Schnitt in E und f eine differenzierbare Funktion bezeichnet.

Bemerkung: $\nabla(s)$ kann als (lokaler) Schnitt in $T^* \otimes E$ auch als (lokaler) Homomorphismus $T \to E$ aufgefaßt werden. Ist dann X ein (lokales) differenzierbares Vektorfeld, so setzt man

$$\nabla_X(s) = \nabla(s)(X) \ .$$

Damit erhält man für jedes solche Vektorfeld X einen $\mathbb{C}$-linearen Homomorphismus

$$\nabla_X: E \to E$$

mit

$$\nabla_X(fs) = f\nabla_X(s) + X(f)s,$$

wobei wie üblich $X(f) = \langle df,X\rangle$, mit $\langle\ ,\ \rangle$ als dualer Paarung $T^* \otimes T \to C^\infty$, und $X(f)$ die Anwendung der Derivation X auf die Funktion f ist.

Den linearen Zusammenhang ∇ kann man auf die äußere Algebra eindeutig durch die Formel

$$\nabla(\omega \otimes s) = d\omega \otimes s + (-1)^p \omega\wedge\nabla(s)$$

fortsetzen, so daß für jedes $p \geq 0$ ein $\mathbb{C}$-linearer Homomorphismus

$$\nabla : \Lambda^p T^* \otimes E \to \Lambda^{p+1} T^* \otimes E$$

definiert ist. Dabei bezeichnet ω eine (lokale) differenzierbare p-Form auf M und s einen (lokalen) differenzierbaren Schnitt in E. Wenn $\nabla(s) = \sum_k \omega_k \otimes s_k$, so ist $\omega\wedge\nabla(s)$ durch $\omega\wedge\nabla(s) = \sum_k (\omega\wedge\omega_k) \otimes s_k$ definiert.

(1.2) Krümmung eines Zusammenhangs.

Durch Hintereinanderausführung erhält man einen Homomorphismus

$$\nabla^2: E \to \Lambda^2 T^* \otimes E$$

mit

$$\nabla^2(fs) = f\nabla^2(s),$$

wie man leicht verifizieren kann. Faßt man den (lokalen) Schnitt $\nabla^2(s)$ von $\Lambda^2 T^* \otimes E$ als alternierenden Homomorphismus $T \otimes T \to E$ auf, so ergibt sich für zwei (lokale) Vektorfelder X und Y die Formel

$$\nabla^2(s)(X,Y) = \nabla_X\nabla_Y(s) - \nabla_Y\nabla_X(s) - \nabla_{[X,Y]}(s)$$

wie man leicht mit Hilfe der Formeln

$$d\omega(X,Y) = X(<\omega,y>) - Y(<\omega,X>) - <\omega,[X,Y]>$$

sowie

$$\omega\wedge\tilde{\omega}(X,Y) = <\omega,X><\tilde{\omega},Y> - <\omega,Y><\tilde{\omega},X>$$

nachrechnen kann. Deshalb stellt ∇^2 nichts weiter als die Krümmung des Zusammenhangs ∇ dar.

(1.3) Lokale Formen zu Zusammenhang und Krümmung.

Ist ∇ ein linearer Zusammenhang für das Vektorbündel E über M und sind $e_1,\dots,e_r$ differenzierbare Basis-Schnitte von E über der offenen Menge $U \subset M$, so werden diesen durch ∇ eindeutige 1-Formen $\omega_{\alpha\beta}$ über U zugeordnet durch die Darstellung von ∇e_α durch die Basis-Schnitte:

$$\nabla e_\alpha = \sum_\beta \omega_{\alpha\beta} \otimes e_\beta.$$

Die Matrix $\omega = (\omega_{\alpha\beta})_{1\le\alpha,\beta\le r}$ von 1-Formen über U heißt die lokale Zusammenhangsmatrix zur Basis $e_1,\dots,e_r$ von ∇. Analog erhält man die lokale Krümmungsmatrix $\Omega = (\Omega_{\alpha\beta})_{1\le\alpha,\beta\le r}$ durch

$$\nabla^2 e_\alpha = \sum_\beta \Omega_{\alpha\beta} \otimes e_\beta,$$

wo jedes $\Omega_{\alpha\beta}$ eine 2-Form über U ist. Man berechnet dann sofort:

$$\Omega_{\alpha\beta} = d\omega_{\alpha\beta} - \sum_\gamma \omega_{\alpha\gamma}\wedge\omega_{\gamma\beta}$$

$$d\Omega_{\alpha\beta} = \sum_\gamma \omega_{\alpha\beta}\wedge\Omega_{\gamma\beta} - \sum_\gamma \Omega_{\alpha\gamma}\wedge\omega_{\gamma\beta}.$$

In Matrizenschreibweise (mit $\wedge$ als Multiplikation) ergeben sich dann:

$$\Omega = d\omega - \omega\wedge\omega,$$

$$d\Omega = \omega\wedge\Omega - \Omega\wedge\omega. \text{ (Bianchi-Identität)}$$

Die lokalen Zusammenhangs- bzw. Krümmungsmatrizen transformieren sich bei Basiswechsel wie folgt. Sind $e_1,\ldots,e_r$ bzw. $e_1',\ldots,e_r'$ zwei Basen von E über U, und bezeichnen ω,Ω bzw. ω',Ω' die zugehörigen Matrizen und ist die invertierbare, differenzierbare Matrix $g = (g_{\alpha\beta})$ durch

$$e_\alpha' = \sum_\beta g_{\alpha\beta} e_\beta$$

bestimmt, so berechnet man leicht:

$$\omega' = dg\, g^{-1} + g\omega g^{-1}$$

$$\Omega' = g\Omega g^{-1} .$$

Bemerkung: In der Physik wird ω als Potential und Ω als zugehöriges Feld interpretiert bezüglich des Meßsystems $e_1,\ldots,e_r$. Eine wie hier angegebene Transformation ist dann eine Eichtransformation.

(1.4) Chernklassen und Satz von Gauß-Bonnet.

Dem differenzierbaren Vektorbündel E über M mit Faser $\mathbb{C}^r$ sind die Chernklassen $c_\nu = c_\nu(E) \in H^{2\nu}(M,\mathbb{Z})$ zugeordnet, $\nu = 1,\ldots,r$, [12]. Diese können nach Chern-Weil als de Rham Cohomologieklassen wie folgt interpretiert werden. Wir beschränken uns dabei auf den Fall $r = 2$. Ist der Zusammenhang ∇ auf E gegeben und Ω eine lokale Krümmungsmatrix zu ∇ , so definieren wir die Formen

$$\sigma_1(\nabla) = \frac{1}{2\pi i} \operatorname{Spur}(\Omega) = \frac{1}{2\pi i}(\Omega_{11} + \Omega_{22})$$

und

$$\sigma_2(\nabla) = \frac{1}{(2\pi i)^2} \det(\Omega) = -\frac{1}{4\pi^2}(\Omega_{11}\wedge\Omega_{22} - \Omega_{21}\wedge\Omega_{12}) .$$

Diese Formen sind, wie man sofort verifizieren kann, invariant unter Basis-Transformationen, 1.3, und stellen somit unabhängig von den lokalen Basen von E globale Formen über M dar. Sie sind Invarianten des Paares (E,∇). Die Form $\sigma_1(\nabla)$ ist eine 2-Form und $\sigma_2(\nabla)$ eine 4-Form. Es gilt, [17] :

(i) $d\sigma_\nu(\nabla) = 0.$

Deswegen definieren $\sigma_\nu(\nabla)$ de Rham-Cohomologieklassen $[\sigma_\nu(\nabla)] \in H^{2\nu}(M,\mathbb{C})$. Ferner gilt:

(ii) Sind ∇ und ∇' zwei lineare Zusammenhänge auf E, so ist $[\sigma_\nu(\nabla)] = [\sigma_\nu(\nabla')]$.

Die Klassen $[\sigma_\nu(\nabla)]$ sind somit Invarianten des Bündels E selbst. Aus dem klassischen Satz von Gauß-Bonnet folgt sogar, [17] :

(iii) Die Klasse $[\sigma_\nu(\nabla)]$ ist das Bild unter dem kanonischen Homomorphismus $H^{2\nu}(M,\mathbb{Z}) \to H^{2\nu}(M,\mathbb{C})$ der Chernklasse $c_\nu(E)$.

Diese Tatsache kann wie folgt interpretiert werden. Ist $\xi \in Z_{2\nu}(M,\mathbb{Z})$ ein differenzierbarer 2ν-Zyklus und $\tilde{c}_\nu : C_{2\nu}(M,\mathbb{Z}) \to \mathbb{Z}$ ein Repräsentant der Chernklasse $c_\nu(E)$, so gilt

$$\int_\xi \sigma_\nu(\nabla) = \tilde{c}_\nu(\xi) \in \mathbb{Z} .$$

Insbesondere erhält man

<u>(1.5) Satz:</u> Ist E ein differenzierbares Vektorbündel vom Rang 2 über der Sphäre S^4 mit dem Zusammenhang ∇ , so ist

$$\int_{S^4} \sigma_2(\nabla) = \tilde{c}_2(S^4) =: c_2 \in \mathbb{Z} .$$

Denn S^4 selbst kann als erzeugender Zyklus für die Homologie $H_4(S^4,\mathbb{Z}) \cong \mathbb{Z}$ aufgefaßt werden. Vermöge dieses kanonischen Zyklus erhält man auch einen Isomorphismus $H^4(S^4,\mathbb{Z}) \cong \mathbb{Z}$ durch den $c_2(E) = c_2$ mit der ganzen Zahl $\tilde{c}_2(S^4)$ identifiziert wird.

Bemerkung: Die Zahl $c_2(E)$ ist die einzige topologische (differenzierbare) Invariante des Bündels E über S^4. Die topologischen Vektorbündel über S^4 werden vollkommen durch diese Zahlen klassifiziert, [21] . Man beachte, daß hier $c_1(E) = 0$ wegen $H^2(S^4,\mathbb{Z}) = 0$.

(1.6) SU(2)-Vektorbündel und SU(2)-verträgliche Zusammenhänge. Jedes differenzierbare komplexe Vektorbündel E über M besitzt eine hermitische Metrik $< , > : E \times E \to C^\infty$, so daß für (lokale) differenzierbare Schnitte s,t in E $\langle s,t\rangle$ eine differenzierbare Funktion ist mit $\langle s,t\rangle(x) = \langle s(x),t(x)\rangle$. Eine solche Metrik kann auf die äußere Algebra mit Koeffizienten in E fortgesetzt werden vermöge

$$\langle \omega \otimes s, \omega' \otimes s'\rangle = \langle s,s'\rangle\, \omega \wedge \overline{\omega}^{\prime T}$$

als Abbildung

$$(\Lambda^p T^* \otimes E) \times (\Lambda^q T^* \otimes E) \to \Lambda^{p+q} T^* .$$

Ferner existiert stets ein mit der Metrik verträglicher Zusammenhang ∇ auf E, so daß für (lokale) differenzierbare Schnitte s,t gilt, [24] ,

$$d\langle s,t\rangle = \langle \nabla s,t\rangle + \langle s,\nabla t\rangle \; .$$

Ist dann $e_1,\dots,e_r$ eine lokale und bezüglich der Metrik orthonormale Basis von E über der offenen Menge U und ω die Zusammenhangsmatrix von ∇ zu dieser Basis, so folgt sofort aus obiger Beziehung, daß

$$\omega + \overline{\omega}^t = 0 \; .$$

Ist darüberhinaus E ein SU(2)-Bündel, so kann man ∇ so wählen, daß die lokalen Matrizen ω "Werte" in $\mathfrak{su}(2)$, der Liealgebra von SU(2) haben, d.h. $\omega + \overline{\omega}^t = 0$ und Spur $\omega = 0$, sofern die Basis $e_1,\dots,e_r$ mit der SU(2)-Struktur von E verträglich sind, [14] . ∇ soll dann SU(2) oder $\mathfrak{su}(2)$-Zusammenhang heißen.

(1.7) Explizite Ausdrücke von SU(2)-Zusammenhängen über S^4 oder $\mathbb{R}^4$.

Im folgenden wollen wir annehmen, daß E ein differenzierbares

SU(2)-Bündel über S^4 ist mit einer hermitischen Metrik $\langle\ ,\ \rangle$, und daß ∇ ein mit dieser Metrik verträglicher SU(2)-Zusammenhang auf E ist. Dann haben ω bzw. Ω bei einer SU(2)-Basis Werte in $\mathfrak{su}(2)$.

Ist $U = S^4 \setminus \{\text{Punkt}\}$ eine Karte von S^4 mit $U \cong \mathbb{R}^4$ und mit den Koordinaten $t_1, \ldots, t_4$, etwa durch stereographische Projektion, so hat E über U eine orthonormale SU(2)-Basis. Die dazu gehörende Zusammenhangsmatrix ω schreibt sich dann in der Form

$$\omega = \sum_{\nu=1}^{4} A_\nu \, dt_\nu$$

mit differenzierbaren Matrixfunktionen $A_\nu : \mathbb{R}^4 \to \mathfrak{su}(2)$. Schreibt man analog für die Krümmungsmatrix

$$\Omega = \sum_{\mu<\nu} F_{\mu\nu} \, dt_\mu \wedge dt_\nu$$

mit differenzierbaren Matrixfunktionen $F_{\mu\nu} : \mathbb{R}^4 \to \mathfrak{su}(2)$, so folgt sofort aus $\Omega = d\omega - \omega\wedge\omega$, daß

$$F_{\mu\nu} = \frac{\partial A_\nu}{\partial t_\mu} - \frac{\partial A_\mu}{\partial t_\nu} - [A_\mu, A_\nu] \ .$$

Da $F_{\mu\nu}$ Werte in $\mathfrak{su}(2)$ hat, folgt ferner für die Chernform $\sigma_2(\nabla)$ über U

(1.7.1) $$8\pi^2\sigma_2(\nabla) = \text{Spur } \Omega\wedge\Omega$$
$$= 2 \text{ Spur } (F_{12}F_{34} - F_{13}F_{24} + F_{23}F_{14}) dt_1 \wedge \ldots \wedge dt_4 .$$

Daraus ergibt sich die Formel

$$c_2 = \int_{S^4} \sigma_2(\nabla) = \frac{1}{4\pi^2} \int_{\mathbb{R}^4} \text{Spur}(F_{12}F_{34} - F_{13}F_{24} + F_{23}F_{14}) dt_1 \ldots dt_4 \ .$$

(1.7.2) Das Wirkungsintegral des Feldes ∇^2 wird definiert durch

$$W(\nabla^2) = \|\Omega\|^2 := -\int_{S^4} \text{Spur } \Omega\wedge *\Omega ,$$

wobei die Form Spur $\Omega\wedge *\Omega$ ebenso wie Spur $\Omega\wedge\Omega$ unabhängig von den lokalen Basen des Bündels ist, und wobei $*\Omega$ wie in (1.8) erklärt ist. Man berechnet wie vorher, wenn $F_{\mu\nu} = (f_{\mu\nu,\alpha\beta})_{\alpha,\beta}$ ist, daß

$$\|\Omega\|^2 = \int_{\mathbb{R}^4} \text{Spur} \sum_{\mu<\nu} F_{\mu\nu}\bar{F}^t_{\mu\nu}\, dt_1\ldots dt_4$$

$$= \int_{\mathbb{R}^4} \sum_{\substack{\mu<\nu\\ \alpha,\beta}} |f_{\mu\nu\alpha\beta}|^2\, dt_1\ldots dt_4 .$$

(1.8) Auf der Riemannschen Mannigfaltigkeit S^4 ist der Operator $*$ für die äußeren Differentialformen wie üblich erklärt. Insbesondere hat man den Operator

$$* : \Lambda^2T^* \to \Lambda^2T^* .$$

Ist $U = S^4\setminus\{\text{Punkt}\}$ eine Karte mit $U \cong \mathbb{R}^4$ und den Koordinaten $t_1,\ldots,t_4$, etwa durch stereographische Projektion, so gilt

$$* \, dt_1\wedge dt_2 = dt_3\wedge dt_4$$

$$* \, dt_1\wedge dt_3 = -dt_2\wedge dt_4$$

$$* \, dt_2\wedge dt_3 = dt_1\wedge dt_4 .$$

Eine lokale 2-Form Ω auf S^4 (mit Werten in $\mathfrak{su}(2)$) heißt selbstdual, falls $*\Omega = \Omega$. Hat Ω bezüglich der Koordinaten $t_1,\ldots,t_4$ die Darstellung

$$\Omega = \sum_{\mu<\nu} F_{\mu\nu}\, dt_\mu\wedge dt_\nu ,$$

so ist $*\Omega = \Omega$ äquivalent zu den Bedingungen

$$F_{12} = F_{34} \, , \, F_{13} = -F_{24} \, , \, F_{23} = F_{14} \, .$$

(1.9) Die Euklidischen Yang-Mills Gleichungen für $su(2)$-Potentiale. Sei E über S^4 ein differenzierbares SU(2)-Bündel und ∇ ein SU(2)-Zusammenhang auf E, der mit einer hermitischen Metrik auf E verträglich ist. Durch Variation des Wirkungsintegrals $W(\nabla^2)$ erhält man dafür, daß $W(\nabla^2)$ stationär wird, die Yang-Mills Bedingungen:

(YM) Für festes μ gilt

$$\sum_{\lambda<\mu} \frac{\partial F_{\lambda\mu}}{\partial t_\lambda} - \sum_{\mu<\lambda} \frac{\partial F_{\mu\lambda}}{\partial t_\lambda} = \sum_{\lambda<\mu} [A_\lambda, F_{\lambda\mu}] - \sum_{\mu<\lambda} [A_\lambda, F_{\mu\lambda}] \, ,$$

wobei

$$F_{\mu\nu} = \frac{\partial A_\nu}{\partial t_\mu} - \frac{\partial A_\mu}{\partial t_\nu} - [A_\mu, A_\nu] \, .$$

Diese Bedingungen sind äquivalent zu der Gleichung

$$d(*\Omega) = \omega\wedge(*\Omega) - (*\Omega)\wedge\omega \, .$$

Ist ∇^2 selbstdual, d.h. $*\Omega = \Omega$, so ist diese Gleichung weiter nichts als die Bianchi-Identität. Insbesondere erfüllen selbstduale Felder die Yang-Mills Gleichungen. Ebenso erfüllen anti-selbstduale Felder mit $*\Omega = -\Omega$ die Yang-Mills Gleichungen.

Man kann nun die Krümmungsmatrix Ω zu ∇ über $\mathbb{R}^4$ aufspalten in einen selbstdualen und einen anti-selbstdualen Anteil:

$$\Omega^+ = \frac{1}{2}(\Omega + *\Omega) \qquad , \; \Omega^- = \frac{1}{2}(\Omega - *\Omega) \quad ,$$

so daß gilt:

$$\Omega = \Omega^{+} + \Omega^{-} \quad , \quad *\Omega^{+} = \Omega^{+}$$

$$\Omega^{+} \wedge \Omega^{-} = 0 \quad , \quad *\Omega^{-} = -\Omega^{-} \quad .$$

Diese Zerlegung ist invariant unter Koordinaten- und Basistransformationen und liefert eine entsprechende Zerlegung von ∇ in $\nabla = \nabla_{+} + \nabla_{-}$, so daß $\Omega^{\pm}$ die Krümmungsmatrix von $\nabla_{\pm}$ ist. Man berechnet ferner, daß

$$\|\Omega\|^2 = \|\Omega^{+}\|^2 + \|\Omega^{-}\|^2$$

und

$$8\pi^2 c_2 = \|\Omega^{+}\|^2 - \|\Omega^{-}\|^2$$

ist. Insbesondere folgt aus diesen Gleichungen weiter

$$\|\Omega\|^2 \geq 8\pi^2 |c_2| .$$

In dieser Abschätzung wird der Wert $8\pi^2|c_2|$ genau dann angenommen, wenn

$$\Omega^{-} = 0 \quad \text{falls} \quad c_2 > 0$$

$$\Omega = 0 \quad \text{falls} \quad c_2 = 0$$

$$\Omega^{+} = 0 \quad \text{falls} \quad c_2 < 0 \; .$$

Der erste und dritte Fall können untereinander vertauscht werden durch Veränderung der Orientierung von S^4, während der mittlere der triviale Fall ist. Um also ein absolutes Minimum von $\|\Omega\|^2$ zu erreichen, betrachten wir im folgenden SU(2)-Bündel E auf S^4 mit $c_2>0$, versehen mit Metrik-verträglichen SU(2)-Zusammenhängen ∇ , und setzen voraus, daß $\nabla_{-} = 0$, d.h. für jede lokale Krümmungsmatrix Ω von ∇ gilt $*\Omega = \Omega$.

(1.10) Bemerkung über die Fortsetzung von euklidischen Yang-Mills Feldern von $\mathbb{R}^4$ auf S^4 .

Ist auf $\mathbb{R}^4$ ein Potential $\omega = \sum_\mu A_\mu \, dt_\mu$ gegeben mit differenzierbaren $A_\mu : \mathbb{R}^4 \to \mathfrak{su}(2)$, so kann man ω als Zusammenhangsmatrix in dem trivialen Bündel $\mathbb{R}^4 \times \mathbb{C}^2$ auffassen, und das Feld $\Omega = d\omega - \omega\wedge\omega$ des Potentials, 1.7, als Krümmungsmatrix. Verlangt man, daß das Wirkungsintegral

$$\|\Omega\|^2 = -\int_{R^4} \operatorname{Spur}(\Omega\wedge *\Omega)$$

endlich ist, so folgt daraus zunächst nicht, daß

$$\frac{1}{8\pi^2} \int_{\mathbb{R}^4} \operatorname{Spur}(\Omega\wedge\Omega)$$

eine ganze Zahl ist. Wenn jedoch ω die lokale Formenmatrix über $\mathbb{R}^4$ eines Zusammenhangs ∇ in einem Bündel E über S^4 ist, so ist dieses Integral gleich $c_2(E) \in \mathbb{Z}$, 1.5, 1.7.1. Die Bedingung dafür lautet: Es gibt eine in einem Bereich $\{|t|>\rho\}$, $\rho>>0$, definierte differenzierbare SU(2)-Matrix g, so daß sich die transformierte Formenmatrix

$$\omega_\infty := g^{-1}\omega g - g^{-1}dg$$

in ∞ glatt verhält. Dann definiert nämlich $g_{12} = g^t$ einen Cozyklus für ein differenzierbares SU(2)-Bündel E über S^4, so daß ω, ω_∞ die lokalen Matrizen eines Zusammenhangs in E sind. Ist dies möglich, so ist E bis auf Isomorphie eindeutig bestimmt, da

$$c_2(E) = \frac{1}{8\pi^2} \int_{\mathbb{R}^4} \operatorname{Spur}(\Omega\wedge\Omega)$$

das Bündel charakterisiert, vgl. Bemerkung nach 1.5. Nun kann man wegen $\|\Omega\|^2 < \infty$ erwarten, daß für eine passende solche Matrix g die Form $dg\, g^{-1}$ die Form ω approximiert für $t \to \infty$, so daß

$\omega_\infty = g^{-1}(\omega - dg\, g^{-1})g$ konvergiert für $t \to \infty$. Es scheint dann plausibel, daß man bei geeigneter Wahl von g auch erwarten kann, daß ω_∞ im Punkt ∞ differenzierbar ist.

Zusatz nach der Korrektur: Inzwischen wurde von K. K. Uhlenbeck [25] gezeigt, daß sich alle euklidischen Yang-Mills-Felder in der angegebenen Weise nach S^4 fortsetzen.

§ 2 Koordinaten und Faserung $\mathbb{P}_3(\mathbb{C}) \to S^4$.

Einer Geraden im $\mathbb{P}_3$ durch die Punkte x,y mit den homogenen Koordinaten $x_o,\dots,x_3$ bzw. $y_o,\dots,y_3$ ordnen wir die Plücker-koordinaten

$$p_{ij} = x_i y_j - x_j y_i$$

zu, $i<j$. Diese sind bis auf einen gemeinsamen Faktor $\lambda \neq 0$ durch die Gerade eindeutig bestimmt und definieren somit einen Punkt $p = p(x,y) \in \mathbb{P}_5$ mit den homogenen Koordinaten $p_{o1},\dots,p_{23}$. Außerdem erfüllen die p_{ij} dann die Relation

$$p_{o1}p_{23} - p_{o2}p_{13} + p_{o3}p_{12} = 0 .$$

Bekanntlich erhält man durch diese Zuordnung eine Bijektion von der Menge $\text{Graß}_1(\mathbb{P}_3)$ aller komplexen projektiven Geraden des $\mathbb{P}_3$ auf die "Kleinsche" Quadrik

$$Q = \{p \in \mathbb{P}_5 \mid p_{o1}p_{23} - p_{o2}p_{13} + p_{o3}p_{12} = 0\}.$$

Mit L_p bezeichnen wir die durch $p \in Q$ bestimmte Gerade des $\mathbb{P}_3$. Die Bedingung $z \in L_p$ für einen Punkt $z \in \mathbb{P}_3$ ist gleichbedeutend mit dem Erfülltsein der Relationen

$$\begin{aligned}
p_{12}z_o - p_{o2}z_1 + p_{o1}z_2 &= 0\\
p_{13}z_o - p_{o3}z_1 + p_{o1}z_3 &= 0\\
p_{23}z_o - p_{o3}z_2 + p_{o2}z_3 &= 0\\
p_{23}z_1 - p_{13}z_2 + p_{12}z_3 &= 0
\end{aligned} \qquad (*)$$

(2.1) Transformiert man die Koordinaten p_{ij} des $\mathbb{P}_5$ in Koordinaten x_i mit

$$x_o = p_{o1} + p_{23} \qquad x_2 = p_{13} + p_{o2} \qquad x_4 = p_{o3} - p_{12}$$

$$x_1 = p_{o1} - p_{23} \qquad x_3 = i(p_{13} - p_{o2}) \quad x_5 = i(p_{o3} + p_{12}) ,$$

so folgt sofort

$$x_o^2 - (x_1^2 + \ldots + x_5^2) = 4(p_{o1}p_{23} - p_{o2}p_{13} + p_{o3}p_{12}) .$$

Die Menge der $x \in Q$ mit $x_i = \overline{x}_i$ kann dann mit der Sphäre S^4 identifiziert werden, so daß Q die Komplexifizierung der reellen Mannigfaltigkeit S^4 ist.

(2.2) Sei $\sigma: \mathbb{P}_3 \to \mathbb{P}_3$ definiert durch die Zuordnung

$$(z_o, z_1, z_2, z_3) \longmapsto (-\overline{z}_1, \overline{z}_o, -\overline{z}_3, \overline{z}_2)$$

Dann ist σ ein Diffeomorphismus mit $\sigma^2 = \text{id}$ und ohne Fixpunkte. Folgende Bedingungen sind nun äquivalent für einen Punkt $p \in Q$:

(1) $p \in S^4$

(2) $p_{o1} = \overline{p}_{o1}$, $p_{o2} = \overline{p}_{13}$, $p_{o3} = -\overline{p}_{12}$, $p_{23} = \overline{p}_{23}$

(3) $\sigma(L_p) = L_p$

(4) Mit $z \in L_p$ ist auch $\sigma z \in L_p$.

Eine Gerade p bzw. L_p mit diesen Eigenschaften soll fortan reelle Gerade heißen.

Beweis: (1) <=> (2) folgt sofort aus der Transformation 2.1.
(3) <=> (4) folgt aus $z \neq \sigma z$ und $\sigma^2 z = z$.

(4) => (2) ergibt sich sofort durch Berechnung der Plücker-koordinaten von z und σz. Wenn (2) erfüllt ist, so folgt (4) aus den Relationen (*).

(2.3) Ordnet man einem Punkt $z \in \mathbb{P}_3$ die reelle Gerade durch z und σz bzw. den Punkt $p = p(z) \in S$ mit den Plückerkoordinaten $p_{ij}(z)$ von z und σz zu, so erhält man eine Abbildung

$$\mathbb{P}_3 \xrightarrow{\pi} S^4 ,$$

die in Koordinaten durch $p_{ij} = p_{ij}(z)$ mit

$$p_{o1} = |z_o|^2 + |z_1|^2$$

$$p_{o2} = -z_o\bar{z}_3 + z_2\bar{z}_1$$

$$p_{12} = -z_1\bar{z}_3 - z_2\bar{z}_o$$

$$p_{o3} = z_o\bar{z}_2 + z_3\bar{z}_1$$

$$p_{13} = z_1\bar{z}_2 - z_3\bar{z}_o$$

$$p_{23} = |z_2|^2 + |z_3|^2$$

gegeben wird. Die Fasern von π sind dann gerade die reellen Geraden. Man kann dann zeigen, daß $\mathbb{P}_3 \xrightarrow[\pi]{} S^4$ sogar ein differenzierbares Faserbündel mit Faser $\mathbb{P}_1$ ist.

(2.4) Die Mannigfaltigkeit Q besitzt die Karten $W_{ij} = Q \cap \{p_{ij} \neq 0\}$, $i<j$. Auf W_{o1} etwa erhält man die lokalen komplexen Koordinaten $p^*_{o2}, p^*_{o3}, p^*_{12}, p^*_{13}$ mit $p^*_{ij} = \dfrac{p_{ij}}{p_{o1}}$.

Die reelle Mannigfaltigkeit $S^4 \subset Q$ wird nun bereits von

$$V_1 = S^4 \cap \{p_{01} \neq 0\} \quad \text{und} \quad V_2 = S^4 \cap \{p_{23} \neq 0\}$$

überdeckt. Bezeichnet man mit $U_i \subset \mathbb{P}_3$ die Karten $\{z_i \neq 0\}$, so folgt sofort, daß

$$\pi^{-1}V_1 = U_0 \cup U_1 \qquad \text{und} \qquad \pi^{-1}V_2 = U_2 \cup U_3 \ .$$

Ferner ergibt sich in den Koordinaten x von 2.1, daß für $x \in S^4$ stets $x_0 \neq 0$ ist. In den inhomogenen Koordinaten $y_1 = \frac{x_1}{x_0}, \ldots, y_5 = \frac{x_5}{x_0}$ der Karte $\{x_0 \neq 0\}$ wird dann S^4 gerade als die übliche Sphäre des $\mathbb{R}^5$ beschrieben. In diesen Koordinaten ist dann

$$V_1 = \{y_1 \neq -1\} \qquad \text{und} \qquad V_2 = \{y_1 \neq 1\},$$

wie man leicht verifiziert. Führt man nun die lokalen reellen Koordinaten $t_1, \ldots, t_4$ auf V_1 durch stereographische Projektion ein, so gilt etwa

$$t_1 = \frac{y_2}{1+y_1}\ , \ t_2 = \frac{y_3}{1+y_1}\ , \ t_3 = \frac{y_4}{1+y_1}\ , \ t_4 = \frac{y_5}{1+y_1}\ .$$

Die entsprechenden Koordinaten $s_1, \ldots, s_4$ in V_2 sind dann durch $s_1 = \frac{y_2}{1-y_1}, \ldots, s_4 = \frac{y_5}{1-y_1}$ gegeben.

(2.5) Die Abbildung $\pi: \mathbb{P}_3 \to S^4$ hat in den Koordinaten z_1^*, z_2^*, z_3^* mit $z_\nu^* = \frac{z_\nu}{z_0}$ von U_0 und $t_1, \ldots, t_4$ von V_1 die Form

$$t_1 = \tfrac{1}{2}(p_{13}^* + p_{02}^*) \qquad t_2 = \tfrac{i}{2}(p_{13}^* - p_{02}^*)$$

$$t_3 = \tfrac{1}{2}(p_{03}^* - p_{12}^*) \qquad t_4 = \tfrac{i}{2}(p_{03}^* + p_{12}^*)\ ,$$

wobei $p_{ij}^* = \frac{p_{ij}(z)}{p_{01}(z)}$ Funktionen in den z_ν^* und $\overline{z_\nu^*}$ sind.

Denn es ist etwa $p_{o2}^* = \dfrac{-\overline{z_3^*}+z_2^*\overline{z_1^*}}{1+z_1^*\overline{z_1^*}}$, $p_{13}^* = \overline{p_{o2}^*}$, und

$$t_1 = \frac{y_2}{1+y_1} = \frac{x_2}{x_o+x_1} = \frac{p_{13}+p_{o2}}{2p_{o1}} = \frac{1}{2}(p_{13}^*+p_{o2}^*) \ .$$

Insbesondere folgt, daß π eine differenzierbare Abbildung ist. Als differenzierbare Funktionen auf U_o besitzen die Funktionen t_ν auch die Ableitungsformen ∂t_ν, $\bar\partial t_\nu$ vom Typ (1,0) bzw. (0,1) mit $dt_\nu = \partial t_\nu + \bar\partial t_\nu$. Es gelten sogar die Identitäten:

$$\text{(2.6)} \quad \bar\partial t_1 \wedge \bar\partial t_2 + \bar\partial t_3 \wedge \bar\partial t_4 = 0$$

$$\bar\partial t_1 \wedge \bar\partial t_3 - \bar\partial t_2 \wedge \bar\partial t_3 = 0$$

$$\bar\partial t_2 \wedge \bar\partial t_3 + \bar\partial t_1 \wedge \bar\partial t_4 = 0 \ .$$

Den Beweis führt man am einfachsten mit den Relationen $\bar\partial p_{12}^* = z_1^*\bar\partial p_{o2}^*$ und $\bar\partial p_{13}^* = z_1^*\bar\partial p_{o3}^*$, die aus den Identitäten

$$z_2^* = p_{o2}^* z_1^* - p_{12}^* \quad \text{und} \quad z_3^* = p_{o3}^* z_1^* - p_{13}^* \ ,$$

(wegen $z \in L_p$ bzw. (*)) folgen.

Da die t_ν reellwertige Funktionen sind, gelten die analogen Identitäten auch für ∂t_ν. Ebenso ergeben sich analoge Identitäten für die Abbildung π in anderen Karten. Daraus ergibt sich schließlich das

(2.7) Lemma: Ist Ω eine (lokale) differenzierbare 2-Form über S^4, so ist $*\Omega = \Omega$ genau dann erfüllt, wenn $\pi^*\Omega$ eine Form vom Typ (1,1) ist.

Der Beweis ergibt sich durch direktes Ausrechnen der Formen in lokalen Koordinaten und Benützen der Identitäten 2.6.

(2.8) Bemerkung: Die Identitäten 2.6 hängen von der Orientierung ab. Bei Änderung der Orientierung ist in 2.7 sinngemäß $*\Omega = -\Omega$ zu verwenden.

§ 3 Penrose-Transformation und Satz von Atiyah-Ward.

Im folgenden werden die Bezeichnungen und Koordinaten wie in § 2 benutzt. Mit Z wird die Fahnenmannigfaltigkeit

$$Z = \{(z,p) \in \mathbb{P}_3 \times Q \mid z \in L_p\}$$

bezeichnet. Dabei ist $z \in L_p$ durch die Relationen (*) von § 2 bzw. durch $P_{1,1}Z_o = 0$ in der Notation von 4.2 gekennzeichnet. Mit p bzw. q seien die kanonischen holomorphen Projektionen von Z nach $\mathbb{P}_3$ bzw. Q bezeichnet. Dabei ist $p : q^{-1}S^4 \to \mathbb{P}_3$ ein Diffeomorphismus und man erhält die Diagramme

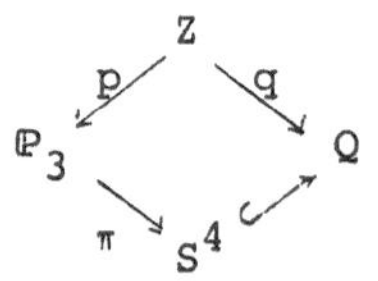

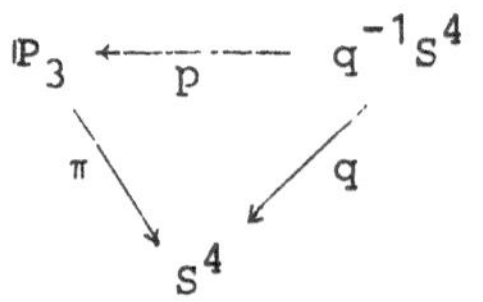

(3.1) Ist F ein differenzierbares komplexes Vektorbündel auf S^4, so ist $\pi^*F = p_*q^*F$ ein differenzierbares Vektorbündel auf $\mathbb{P}_3$. Wenn F Einschränkung eines holomorphen Vektorbündels $\tilde{F}$ über Q ist, so ist $p_*q^*\tilde{F}$ als kohärente Garbe lokal-frei auf $\mathbb{P}_3 \setminus \Sigma$, wo $\Sigma = \{z \in \mathbb{P}_3 \mid \tilde{F}|G_z \text{ nicht-trivial}\}$ und wo G_z die Menge der $p \in Q$ mit $z \in L_p$. Bedingungen, unter denen π^*F ein holomorphes Vektorbündel auch ohne eine Komplexifizierung $\tilde{F}$ ist, sind von Atiyah-Ward [4] angegeben worden. Diese sollen im folgenden erläutert werden.

(3.2) Ist umgekehrt E ein holomorphes Vektorbündel auf $\mathbb{P}_3$ und $\Sigma = \{p \in Q \mid E|L_p \text{ nicht-trivial}\}$, so ist die kohärente Garbe

$F = q_* p^* E$ lokal-frei auf $Q \setminus \Sigma$ und die Faser F_p des Vektorbündels F ist dann für $p \in Q \setminus \Sigma$ durch $F_p = \Gamma(L_p, E|L_p)$ gegeben.

Beweisskizze: $Z \xrightarrow[q]{} Q$ ist ein Faserbündel mit Fasern $L_p \cong \mathbb{P}_1$. Ist $p \in Q \setminus \Sigma$, so gibt es eine Umgebung $U(p) \subset Q$ mit $q^{-1}(U) \cong U \times \mathbb{P}_1$. Aus der Deformationstheorie folgt, daß $p^* E$ dann auf $q^{-1}(U)$ trivial ist für U hinreichend klein, so daß $q_* p^* E$ über U trivial ist. Es folgt dann leicht, daß $F_p = \Gamma(L_p, E|L_p) \cong \mathbb{C}^r$.

(3.3) Sei nun F ein differenzierbares $SU(2)$-Bündel über S^4 mit hermitischer Metrik und verträglichem Zusammenhang ∇. Wenn $*\nabla = \nabla$, so hat $E = \pi^* F$ eine holomorphe Struktur, so daß $\tilde{\nabla} = \pi^* \nabla$ der mit der gelifteten Metrik verträgliche eindeutige holomorphe Zusammenhang auf E wird. Außerdem ist $E|L_p$ analytisch-trivial für jede reelle Gerade $p \in S^4$. Für die Chernklassen gilt $c_1(E) = 0$, $c_2(E) = c_2(F) > 0$.

Beweis: Die Mengen $\pi^{-1}V_1 = U_0 \cup U_1$, $\pi^{-1}V_2 = U_2 \cup U_3$ seien wie in § 2. Das Bündel F kann durch einen Cozyklus $g: V_1 \cap V_2 \to SU(2)$ gegeben werden und E durch $\pi^* g : \pi^{-1}V_1 \cap \pi^{-1}V_2 \to SU(2)$. Seien $\psi_\nu : E|\pi^{-1}V_\nu \to \pi^{-1}V_\nu \times \mathbb{C}$ die gelifteten Trivialisierungen. Wir setzen $\tilde{g}_{01} = \mathrm{id}$, $\tilde{g}_{23} = \mathrm{id}$ und $\tilde{g}_{ij} = g \circ \pi$ für $i < j$, $(i,j) \neq (0,1), (2,3)$, so daß $(\tilde{g}_{ij})$ der Cozyklus bzgl. der Trivialisierungen ψ ist. E hat eine holomorphe Struktur genau dann, wenn es Trivialisierungen $\varphi_i : E|U_i \to U_i \times \mathbb{C}^2$ gibt mit $g_{ij} = \varphi_i \varphi_j^{-1}$ holomorph. Dann gilt mit $h_i = \varphi_i \psi_\nu^{-1}$, $i = 0,1$ für $\nu = 1$ bzw. $i = 2,3$ für $\nu = 2$

$$g_{ij} = h_i \tilde{g}_{ij} h_j^{-1} \quad . \qquad (*)$$

Gibt es umgekehrt differenzierbare $h_i : U_i \to GL(2,\mathbb{C})$, so daß g_{ij}

holomorph sind, so erhält E eine holomorphe Struktur mit Trivialisierungen den $\varphi_i = h_i\psi_\nu$. Die Bedingung (*) lautet dann

$$\bar\partial\tilde g_{ij} = \tilde g_{ij}h_j^{-1}\bar\partial h_j - h_i^{-1}\bar\partial h_i\tilde g_{ij} \, .$$

Ist nun andererseits ω_ν die Zusammenhangsmatrix von F auf V_ν von ∇, und $\tilde\omega_\nu = \pi^*\omega_\nu$ die von $\pi^*\nabla = \tilde\nabla$ über $\pi^{-1}V_\nu$, so folgt aus

$$\omega_2 \, g^t = dg^t + g^t \, \omega_1$$

$$\bar\partial(g\circ\pi) = g(\pi^*\omega_2^t)'' - (\pi^*\omega_1^t)''g \, ,$$

wo " wie üblich den Anteil vom Typ (0,1) von 1-Formen bezeichnet. Kann man nun die Gleichungen

$$(\pi^*\omega_\nu^t)'' = h_i^{-1}\bar\partial h_i$$

über U_i lösen, so erhält man auf E eine holomorphe Struktur. Die Bedingungen für die Lösbarkeit lauten aber ,[18],[15],[14],

$$\bar\partial(\pi^*\omega_\nu) - (\pi^*\omega_\nu)''\wedge (\pi^*\omega_\nu)'' = 0 \quad ,$$

oder gleichbedeutend $(\pi^*\Omega_\nu)'' = 0$, wenn $\Omega_\nu = d\omega_\nu - \omega_\nu\wedge\omega_\nu$ die zugehörige Krümmungsmatrix ist. Diese Bedingung ist nach 2.7 gerade erfüllt, wenn $*\nabla = \nabla$.

Trägt nun E eine durch solche h_i gegebene holomorphe Struktur, so ist $\tilde\nabla = \pi^*\nabla$ der eindeutige mit der Metrik verträgliche holomorphe Zusammenhang. Denn ist ω_i die Formenmatrix bzgl. der holomorphen Trivialisierung φ_i von $\tilde\nabla$, so gilt nach dem Transformationsgesetz

$$(\pi^*\omega_\nu)h_i^t = dh_i^t + h_i^t\omega_i$$

und wegen obiger Lösung durch h_i folgt daraus $\omega_i'' = 0$.

Da $\pi^*\omega_\nu$ in Richtung der Koordinaten einer Faser $\pi^{-1}(p) = L_p$ verschwinden, $\omega_\nu \circ (\pi | L_p) = 0$, sind die Lösungen h_i holomorph auf L_p. Da $g_{01} = h_o h_1^{-1}$ und $g_{23} = h_2 h_3^{-1}$ und jedes L_p in $U_o \cup U_1$ oder $U_2 \cup U_3$ liegt, folgt, daß $E|L_p$ analytisch trivial ist.

(3.4) Die SU(2)-Struktur auf dem Bündel F über S^4 erlaubt es, eine antilineare Involution $\sigma : F \to F$ invariant zu definieren, indem man für eine Trivialisierung $\sigma : \mathbb{C}^2 \to \mathbb{C}^2$ durch $(z,w) \to (-\bar{w},\bar{z})$ definiert. Dabei ist σ verträglich mit der Metrik, d.h. $\langle \sigma s, \sigma t \rangle = \overline{\langle s,t \rangle}$, und $\sigma^2 = -1$. Vermöge

$$\begin{array}{ccc} F_p & \xrightarrow{\sigma_p} & F_p \\ \Big\downarrow \wr & & \Big\downarrow \wr \\ E_z & \xrightarrow[\sigma_z]{} & E_{\sigma z} \end{array} \qquad p = \pi(z)$$

ist dann für $E = \pi^*F$ eine antilineare Involution $\sigma : E \to \sigma^*E$ definiert, die sogar holomorph von z abhängt, $\sigma^2 = -1$. Somit erhält E durch die SU(2)-Struktur von F eine "symplektische" Struktur σ. Durch $(s,t) = \langle s, -\sigma t \rangle$ ist auf E eine nicht-entartete symplektische Form definiert mit $(\sigma s, \sigma t) = \overline{(s, t)}$. Das holomorphe Bündel E zusammen mit σ bestimmt nun umgekehrt den selbst-dualen Zusammenhang ∇:

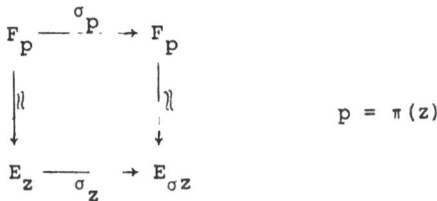

(3.5) Satz von Atiyah-Ward. [4] :

Es besteht eine bijektive Korrespondenz zwischen

(I) den Äquivalenzklassen (unter Eichtransformation) von euklidischen, selbstdualen SU(2)-Yang-Mills Feldern auf S^4 und

(II) den Isomorphieklassen von holomorphen Vektorbündeln E vom Rang 2 auf $\mathbb{P}_3$ mit

(i) $E|L_p$ ist analytisch trivial für jede reelle Gerade $p \in S^4$.

(ii) E besitzt einen holomorphen Isomorphismus $\sigma: E \to \sigma^*\overline{E}$ mit $\sigma^2 = -1$.

Bemerkungen: 1) Mit E ist auch das Bündel $\sigma^*\overline{E}$ ein holomorphes Bündel, denn ist (g_{ij}) ein Cozyklus von E, so hat $\sigma^*\overline{E}$ den Cozyklus $g_{ij}^{\sigma} = \overline{g}_{\sigma(i)\sigma(j)} \circ \sigma$, vgl. § 5.

2) Aus (i) folgt, daß $c_1 = 0$ und (i), (ii) implizieren weiter, daß E keine holomorphen Schnitte außer O hat, 5.2. Nach [6] ist dann E stabil, d.h. $\mathrm{Hom}(\mathbb{P}_3, E, E) = \mathbb{C}$ und $c_2 > 0$.

3) Die Involution $\sigma: E \to \sigma^*\overline{E}$ ist bis auf eine Konstante c mit $|c| = 1$ eindeutig bestimmt, 5.4.

Definition: Ein holomorphes Vektorbündel der Klasse (II) auf $\mathbb{P}_3$ heißt Instantonenbündel.

Beweis des Satzes von Atiyah-Ward: Die Zuordnung $(F,\nabla) \to (E,\sigma)$ ergibt sich aus 3.3, 3.4. Umgekehrt erhält man (F,∇) aus (E,σ) wie folgt. Wegen (i) und 3.2 existiert ein Vektorbündel F auf S^4 mit $E = \pi^*F$ im differenzierbaren Sinne. Es ist dann

$$F_p = \Gamma(L_p, E|L_p) \cong \mathbb{C}^2 ,$$

wobei Γ den Raum der holomorphen Schnitte bezeichnet.

Das Bündel F kann als U(2)-Bündel angenommen werden. Da $c_1 = 0$, gibt es einen Isomorphismus $\alpha : \Lambda^2 E \to \mathcal{O}$, so daß die Form $(s,t) = \alpha(s\wedge t)$ eine holomorphe nirgends entartete symplektische Bilinearform auf E ist. Diese induziert eine symplektische Form (,) auf F über die induzierte Paarung auf $\Gamma(L_p, E|L_p)$

$$(\ , \ ) : F_p \otimes F_p \to \mathbb{C} .$$

Die Involution σ induziert ebenfalls einen antilinearen Isomorphismus

$$\sigma_p : \Gamma(L_p, E|L_p) \longrightarrow \Gamma(L_p, E|L_p)$$

mit $\sigma_p^2 = -1$ für jedes p. Außerdem induziert $\sigma\wedge\sigma$ vermöge α einen antilinearen Isomorphismus $\mathcal{O} \to \sigma^*\overline{\mathcal{O}}$. Dieser ist bis auf eine Konstante c vom Betrag 1 eindeutig durch die übliche Konjugation bestimmt, d.h. es ist $\sigma^*\overline{\alpha}(\sigma s\wedge\sigma t) = c\overline{\alpha(s\wedge t)}$.
Daraus ergibt sich für die Bilinearform (,) auf F_p das Gesetz

$$(\sigma_p s, \sigma_p t) = c\overline{(s,t)} .$$

Man kann daher auf F_p eine Metrik durch

$$\langle s,t\rangle = \frac{1}{\sqrt{c}}(s,\sigma_p t)$$

definieren, die dann eine hermitische Metrik auf dem Bündel F ergibt. Vermöge der kanonischen Isomorphismen $\Gamma(L_p, E|L_p) \to E_z$ für $z \in L_p$, $p \in S^4$, wird nun eine hermitische Metrik auf E definiert, die konstant entlang der reellen geraden L_p ist.
Es gilt dann stets

$$\langle\sigma s,\sigma t\rangle = \overline{\langle s,t\rangle} \; .$$

Daraus folgt, daß F sogar die Struktur eines SU(2)-Bündels trägt. Denn ändert man durch unitäre Transformation die Trivialisierungen so, daß σ die kanonische Form $\begin{pmatrix} 0 & -1 \\ 1 & 0 \end{pmatrix}$ bekommt, so muß der neue Cozyklus von F in SU(2) liegen. Ist nun $\tilde{\nabla}$ der eindeutige holomorphe und zur Metrik $\langle\;,\;\rangle$ gehörende Zusammenhang auf E, so gibt es einen Zusammenhang ∇ auf F mit $\tilde{\nabla} = \pi^*\nabla$, da die Metrik auf $E|L_p$ konstant ist. Da $\tilde{\nabla}$ ein holomorpher Zusammenhang ist, ist ∇ nach 2.7, 3.3 selbstdual mit Werten in $\mathfrak{su}(2)$.

(3.6) Die Zusammenhangsmatrizen ω_ν von ∇ über V_ν kann man wie folgt berechnen, etwa ω_1 : Man konstruiert einen Schnitt

$p \to s(p) \in \Gamma(L_p, E|L_p) = F_p$ über V_1 mit $s(p) \neq 0$ für jedes p. Setzt man $s_1 = s$ und $s_2 = \sigma s$, so sind s_1 und s_2 orthogonal bezüglich der von σ definierten Metrik, denn $\langle s_1, s_2\rangle = s_1 \wedge \sigma s_2 = s \wedge (-s) = 0$. Insbesondere liefern dann s_1, s_2 eine differenzierbare Trivialisierung von F über V_1 bzw. $\psi_1 : E|U_o \cup U_1 \to U_o \cup U_1 \times \mathbb{C}^2$. Ersetzt man s durch $\frac{s}{\sqrt{\langle s,s\rangle}}$, so sind s_1 und s_2 sogar orthonormal. Sind dann $\varphi_i : E|U_i \to U_i \times \mathbb{C}^2$ irgendwelche holomorphen Trivialisierungen, so erhalten wir etwa

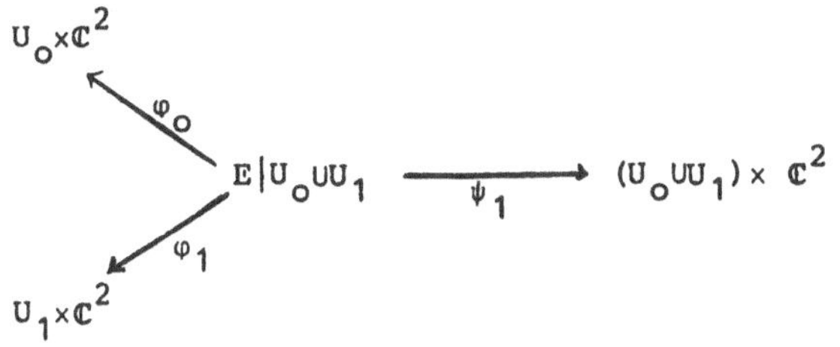

und setzen $h_o = \varphi_o \psi_1^{-1}$ und $h_1 = \varphi_1 \psi_1^{-1}$. Nun setzen wir

$$\tilde{\omega}_1''^{t} = h_o^{-1}\bar{\partial} h_o = h_1^{-1}\bar{\partial} h_1 \quad \text{auf } U_o \cup U_1$$

sowie $\quad \tilde{\omega}_1' = -\overline{\tilde{\omega}_1''^{t}}$.

Diese Formen sind sogar unabhängig von den holomorphen Trivialisierungen, denn ist etwa $h_o = g_o h_o'$ und g_o holomorph, so ist wegen $\bar{\partial} g_o = 0$

$$h_o^{-1}\bar{\partial} h_o = h_o'^{-1} g_o^{-1} g_o \bar{\partial} h_o' = h_o'^{-1}\bar{\partial} h_o' .$$

Durch

$$\tilde{\omega}_1 = \tilde{\omega}_1' + \tilde{\omega}_1''$$

erhält man nun die Zusammenhangsmatrix von $\tilde{\nabla}$ bzgl. s_1, s_2 über V_1, die dann automatisch von der Form $\tilde{\omega}_1 = \pi^* \omega_1$ ist, wo ω_1 die SU(2)-Zusammenhangsmatrix von ∇ auf F ist, mit selbstdualer

Krümmungsmatrix. In § 6 wird dieses Verfahren am Beispiel der Instantonenbündel zur Instantonenzahl $c_2 = 1$ durchgeführt.

§ 4 Darstellung von holomorphen Vektorbündeln über $\mathbb{P}_3$.

In diesem Paragraphen sollen Ergebnisse von [22] ,[23] über die Darstellung von holomorphen Vektorbündeln über $\mathbb{P}_3$ kurz zusammengestellt werden. Ist E ein holomorphes Vektorbündel, so bezeichne $\mathcal{E}$ die lokal-freie Garbe der holomorphen Schnitte von E. Man braucht im wesentlichen nicht zwischen E und $\mathcal{E}$ zu unterscheiden. Mit $\mathcal{O}(d)$ sei das zum Cozyklus $g_{ij} = (\frac{z_j}{z_i})^d$ gehörende Geradenbündel bezeichnet. Wir setzen $p\mathcal{E}$ für die p-fache direkte Summe von $\mathcal{E}$ und $\mathcal{E}(d) = \mathcal{E} \otimes \mathcal{O}(d)$. Der Vektorraum Hom$(\mathbb{P}_3, m\,\mathcal{O}(d), n\,\mathcal{O}(e))$ wird mit dem Raum der m×n-Matrizen aus homogenen Polynomen vom Grad e-d in $z_o,\ldots,z_n$ in der Weise identifiziert, daß der Matrix M der Homomorphismus entspricht, der lokal über $U_i = \{z_i \neq 0\}$ die Zeile f in $f\circ\frac{M}{z_i^{e-d}}$ überführt.

(4.1) Der Koszul-Komplex im $\mathbb{P}_3$ besteht aus der exakten Sequenz

$$0 \longrightarrow \mathcal{O}(-4) \xrightarrow[Z_3]{} 4\,\mathcal{O}(-3) \xrightarrow[Z_2]{} 6\,\mathcal{O}(-2) \xrightarrow[Z_1]{} 4\,\mathcal{O}(-1) \xrightarrow[Z_o]{} \mathcal{O} \to 0\ ,$$

wo die Matrizen Z_s wie in 4.2 als Matrizen vom Typ $A_{o,s}$ definiert sind mit z_ν anstelle der a_ν, so daß Z_s homogen vom Grad 1 ist. Bekanntlich ist Im Z_s die Garbe der holomorphen s-Formen Ω^s.

(4.2) Ist $e_o,\ldots,e_3$ die kanonische Basis des $\mathbb{C}^4$, so hat jedes $a \in \Lambda^{t+1}\mathbb{C}^4$, $-1\leq t\leq 2$, eine eindeutige Darstellung

$$a = \Sigma\, a_{\nu_o\ldots\nu_t}\, e_{\nu_o}\wedge\ldots\wedge e_{\nu_t}\ ,$$

wo über alle Indexfolgen mit $\nu_o<\ldots<\nu_t$ summiert wird.

Mit $a_{t,s} : \Lambda^{t+s+1}\mathbb{C}^4 \to \Lambda^s\mathbb{C}^4$ sei die zu $\xi \to a\wedge\xi$ duale Abbildung bezeichnet. Sind die kanonischen Basen dieser Räume lexikographisch geordnet, so sei $A_{t,s}$ diejenige $\binom{4}{t+s+1}\times\binom{4}{s}$-Matrix, die vermöge $x \to x\cdot A_{t,s}$ die Abbildung $(-1)^{ts}a_{t,s}$ darstellt.
Für t = 0 haben die $A_s = A_{o,s}$ die Form

$$A_o = \begin{pmatrix} a_o \\ a_1 \\ a_2 \\ a_3 \end{pmatrix},\quad A_1 = \begin{pmatrix} -a_1 & a_o & 0 & 0 \\ -a_2 & 0 & a_o & 0 \\ 0 & -a_2 & a_1 & 0 \\ -a_3 & 0 & 0 & a_o \\ 0 & -a_3 & 0 & a_1 \\ 0 & 0 & -a_3 & a_2 \end{pmatrix}\quad A_2 = \begin{pmatrix} a_2 & -a_1 & a_o & 0 & 0 & 0 \\ a_3 & 0 & 0 & -a_1 & a_o & 0 \\ 0 & a_3 & 0 & -a_2 & 0 & a_o \\ 0 & 0 & a_3 & 0 & -a_2 & a_1 \end{pmatrix},$$

$$A_3 = (-a_3 \quad a_2 \quad -a_1 \quad a_o) .$$

Für t = 1 erhalten $A_{1,s}$ die Form

$$A_{1,o} = \begin{pmatrix} a_{o1} \\ a_{o2} \\ a_{12} \\ a_{o3} \\ a_{13} \\ a_{23} \end{pmatrix},\qquad A_{1,1} = \begin{pmatrix} -a_{12} & a_{o2} & -a_{o1} & 0 \\ -a_{13} & a_{o3} & 0 & -a_{o1} \\ -a_{23} & 0 & a_{o3} & -a_{o2} \\ 0 & -a_{23} & a_{13} & -a_{12} \end{pmatrix},$$

$$A_{1,2} = (a_{23} \quad -a_{13} \quad a_{o3} \quad a_{12} \quad -a_{o2} \quad a_{o1}) .$$

Man berechnet dann sofort, daß

$$A_{1,2}A_{1,o} = 2(a_{o1}a_{23} - a_{o2}a_{13} + a_{o3}a_{12})$$

$$\det A_{1,1} = (a_{o1}a_{23} - a_{o2}a_{13} + a_{o3}a_{12})^2 .$$

(4.2.1) Es gilt stets $Z_{t+s+1}A_{t,s} + A_{t,s+1}Z_s = 0$.

(4.3) Aus dem Koszul-Komplex 4.1 folgt leicht, daß die Zeilen von Z_s gerade eine $\mathbb{C}$-Basis von $\Gamma(\mathbb{P}_3,\Omega^s(s+1))$ bilden, so daß etwa $\omega_{01} = (-z_1\ z_o\ 0\ 0),\dots,\omega_{23} = (0\ 0\ -z_3\ z_2)$ eine Basis von $\Gamma\Omega^1(2) \subset 4\Gamma\mathcal{O}(1)$ bilden.

(4.4) Die kanonische Paarung $\Gamma\mathcal{O}(1)\otimes H^p\mathcal{E}(d-1) \to H^p\mathcal{E}(d)$ induziert für jedes $z_i \in \Gamma\mathcal{O}(1)$ eine lineare Abbildung $\xi \to z_i\xi$ von $H^p\mathcal{E}(d-1) \to H^p\mathcal{E}(d)$. Analog bekommt man eine bilineare Abbildung $\Gamma\Omega^1(2)\otimes H^2\mathcal{E}(-3) \to H^2\Omega^1\otimes\mathcal{E}(-1)$. Ist nun $H^1\mathcal{E}(-2) = 0$, so ergibt sich aus der exakten Sequenz $0 \to \Omega^1\otimes\mathcal{E}(-1) \to 4\,\mathcal{E}(-2) \to \mathcal{E}(-1) \to 0$, die man aus dem Koszul-Komplex durch Tensorieren mit $\mathcal{E}(-1)$ erhält, ein Isomorphismus $H^1\mathcal{E}(-1) \xrightarrow[\delta]{} H^2\Omega^1\otimes\mathcal{E}(-1)$. Dadurch erhält man für jedes $\omega_{ij} \in \Gamma\Omega^1(2)$ eine kanonische lineare Abbildung $H^2\mathcal{E}(-3) \to H^1\mathcal{E}(-1)$, die wir mit $\xi \to \omega_{ij}\xi$ bezeichnen.

(4.5) Im folgenden konstruieren wir holomorphe Vektorbündel auf $\mathbb{P}_3$ zu vorgegebenen Dimensionszahlen $\ell_d^i = \dim H^i(\mathbb{P}_3,\mathcal{E}(d))$ für $i = 1,2$. Sei zunächst $e \geq 0$ eine ganze Zahl und

$$\ell = (\ell_{e-1}^1,\dots,\ell_{-1}^1)$$

ein System von natürlichen Zahlen und seien für $-e-3 \leq d \leq -3$ die Zahlen $\ell_d^2 = \ell_{-d-4}^1$. Dann erhalten wir in

$$\mathcal{F} = \bigoplus_{\substack{d\\1\leq i\leq 2}} \ell_d^i\Omega^i(-d)$$

ein holomorphes Vektorbündel auf $\mathbb{P}_3$, dessen Cohomologiegruppen

$H^i(\mathbb{P}_3, \mathcal{F}(d))$ für i = 1,2 gerade die vorgegebene Dimension ℓ_d^i haben bzw. Null sind, falls i = 1 und $d \notin [-1,e-1]$ oder i = 2 und $d \notin [-e-3,-3]$. Das Bündel $\mathcal{F}$ kann als Bild der Matrix $Z^\ell(0)$ dargestellt werden, die man als direkte Summe wie folgt erhält:

Für s = 1,2 sei die Diagonal-Blockmatrix

$$Z_s^{-d-s} = \begin{pmatrix} Z_1 & & 0 \\ & \ddots & \\ 0 & & Z_s \end{pmatrix}$$

aus ℓ_d^s Matrizen Z_s als Homomorphismus

$$\ell_d^s\binom{4}{s+1}\mathcal{O}(-d-s-1) \xrightarrow[Z_s^{-d-s}]{} \ell_d^s\binom{4}{s}\mathcal{O}(-d-s)$$

aufgefaßt. Die Matrix $Z^\ell(0)$ sei dann analog die direkte Summe der Matrizen

$$Z_1^{-e}, Z^{-e+1}, \dots, Z_1^0, Z_2^1, \dots, Z_2^{e+1},$$

die dann als Homomorphismus

$$\mathcal{L}_1 \xrightarrow[Z^\ell(0)]{} \mathcal{L}_0$$

aufgefaßt wird, wobei

$$\mathcal{L}_1 := 6\ell_{e-1}^1\mathcal{O}(-e-1)\oplus\dots\oplus 6\ell_{-1}^1\mathcal{O}(-1)\oplus 4\ell_{-3}^2\mathcal{O}(0)\oplus\dots\oplus 4\ell_{-e-3}^2\mathcal{O}(e)$$

$$\mathcal{L}_0 := 4\ell_{e-1}^1\mathcal{O}(-e)\oplus\dots\oplus 4\ell_{-1}^1\mathcal{O}(0)\oplus 6\ell_{-3}^2\mathcal{O}(1)\oplus\dots\oplus 6\ell_{-e-3}^2\mathcal{O}(e+1).$$

(4.6) Koeffizientensystem A und Matrizen $Z^\ell(A)$.

Das System ℓ von Dimensionszahlen sei wie in 4.5. Zu ℓ betrachten wir nun Koeffizientensysteme A aus Koeffizienten $a_{ij}^{\mu\nu} \in \mathbb{C}$ und ${}^{\varepsilon}a_i^{\mu\nu} \in \mathbb{C}$ die zu folgenden Indices gegeben seien:

(1) $a_{ij}^{\mu\nu}$ zu $0 \le i < j \le 3$, $1 \le \mu,\nu \le k: = \ell_{-1}^{1}$

mit $a_{ij}^{\mu\nu} = a_{ij}^{\nu\mu}$.

(2) ${}^{\varepsilon}a_i^{\mu\nu}$ zu $0 \le i \le 3$, $-e \le \varepsilon \le -1$, $1 \le \mu \le p = \ell_{-\varepsilon-2}^{1}$, $1 \le \nu \le q = \ell_{-\varepsilon-1}^{1}$.

Aus den Koeffizienten vom Typ (1) bilden wir nun die Matrizen $A_{1,s}^{\mu\nu}$ vom Typ $A_{1,s}$ wie in 4.2 und daraus die Matrix

$$A_{1,s}^{o} = \begin{pmatrix} A_{1,s}^{11} & \cdots & A_{1,s}^{1k} \\ \vdots & & \vdots \\ A_{1,s}^{1k} & \cdots & A_{1,s}^{kk} \end{pmatrix} .$$

Aus den Koeffizienten vom Typ (2) bilden wir die Matrizen ${}^{\varepsilon}A_s^{\mu\nu}$ vom Typ $A_s = A_{o,s}$ wie in 4.2 und daraus

$$A_s^{\varepsilon} = \begin{pmatrix} {}^{\varepsilon}A_s^{11} & \cdots & {}^{\varepsilon}A_s^{1q} \\ \vdots & & \vdots \\ {}^{\varepsilon}A_s^{p1} & \cdots & {}^{\varepsilon}A_s^{pq} \end{pmatrix} \qquad A_s^{-\varepsilon} = \begin{pmatrix} {}^{\varepsilon}A_s^{11} & \cdots & {}^{\varepsilon}A_s^{p1} \\ \vdots & & \vdots \\ {}^{\varepsilon}A_s^{1q} & \cdots & {}^{\varepsilon}A_s^{pq} \end{pmatrix} ,$$

so daß $A_s^{-\varepsilon}$ durch Transponierung der Blöcke aus A_s^{ε} entsteht.

Nun wird die Matrix $Z^{\ell}(A)$ wie folgt festgesetzt:

$$Z^{\ell}(A)=\begin{pmatrix} Z_1^{-e} & 0 & \cdots & & & & & 0\\ A_1^{-e} & Z_1^{-e+1} & \ddots & & & & & \vdots\\ 0 & \ddots & \ddots & \ddots & & & & \vdots\\ \vdots & & A_1^{-1} & Z_1^{0} & \ddots & & & \vdots\\ \vdots & & & A^{0}_{1,1} & Z_2^{1} & \ddots & & \vdots\\ \vdots & & & & A_2^{1} & Z_2^{2} & \ddots & \vdots\\ \vdots & & & & & \ddots & Z_2^{e} & 0\\ 0 & \cdots & & & & 0 & A_2^{e} & Z_2^{e+1}\end{pmatrix}$$

Wie $Z^{\ell}(O)$ definiert die Matrix $Z^{\ell}(A)$ einen Homomorphismus

$$\mathcal{L}_1 \xrightarrow[Z^{\ell}(A)]{} \mathcal{L}_o$$

der freien Bündel, dessen Bildgarbe mit $\mathcal{Z}^{\ell}(A)$ bezeichnet sei.

Das Koeffizientensystem A soll nun noch die <u>quadratischen Relationen</u>

(Q) $A^{o}_{1,1}A_o^{-1} = 0$ und $A_1^{\varepsilon}A_o^{\varepsilon-1} = 0$ für $-e<\varepsilon<o$ erfüllen.

(Dann gelten auch die transponierten Bedingungen $A_3^1 A^{o}_{1,1} = 0$ und $A_3^{\varepsilon}A_2^{\varepsilon-1} = 0$ für $1<\varepsilon<e$.)

(4.6.1) Erfüllt das System A die Relationen (Q), so ist die Bildgarbe $\mathcal{Z}^{\ell}(A)$ von $Z^{\ell}(A)$ lokal-frei und es gilt überdies

$$\dim H^i(\mathbb{P}_3, \mathcal{Z}^{\ell}(A)(d)) = \dim H^i(\mathbb{P}_3, \mathcal{Z}^{\ell}(O)(d))$$

für i = 1,2 und alle d, unabhängig von A, vgl. [22].

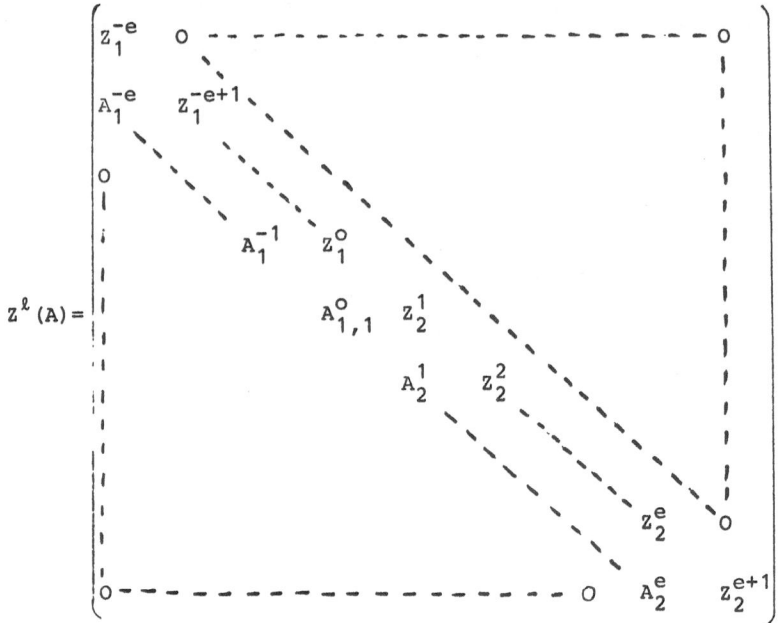

(4.6.2) Setzt man ferner $r_o = \text{rg } A^o_{1,1}$ und $r_\varepsilon = \text{rg } A_1^{-\varepsilon}$ $(=\text{rg } A_2^\varepsilon)$ für $1 \leq \varepsilon \leq e$, so spaltet das Bündel auf in

$$\mathcal{F}^\ell(A) = \mathcal{E}(A) \oplus \mathcal{L}$$

mit

$$\mathcal{L} = r_e\mathcal{O}(-e)\oplus\ldots\oplus r_1\mathcal{O}(-1)\oplus r_o\mathcal{O}(o)\oplus r_1\mathcal{O}(1)\oplus\ldots\oplus r_e\mathcal{O}(e),$$

wobei $\mathcal{E}(A)$ bis auf Isomorphie eindeutig bestimmt ist und für $i = 1,2$ dieselben Cohomologiegruppen H^i besitzt wie $\mathcal{F}^\ell(O)$. Umgekehrt gilt nach [22]:

<u>(4.7) Darstellungssatz:</u> Ist $\mathcal{E}$ ein (nicht spaltendes) holomorphes Vektorbündel vom Rang 2 über $\mathbb{P}_3$ mit $c_1 = 0$ (dann ist $\mathcal{E} \cong \mathcal{E}^*$) und $H^1\mathcal{E}(d) = 0$ für $d \leq -2$, und sind $\ell_d^1 = \dim H^1\mathcal{E}(d)$ für $d \geq -1$, so existiert ein Koeffizientensystem A wie in 4.5, so daß

$$\mathcal{E} \oplus \mathcal{L} = \mathcal{F}^\ell(A),$$

wobei $\mathcal{L}$ wie in 4.6.2 gegeben ist und die Zahlen $r_o, r_1, \ldots, r_e$ Invarianten von $\mathcal{E}$ sind.

(4.7.1) Die Koeffizienten in A zu $\mathcal{E}$ sind durch Auszeichnung von Basiselementen der Vektorräume $H^1\mathcal{E}(d)$ sogar eindeutig bestimmt:

(α) Ist $\xi_1, \ldots, \xi_k$ eine Basis von $H^1\mathcal{E}(-1)$ und $\eta_1, \ldots, \eta_k$ die (bezüglich Serre-Dualität) duale Basis von $H^2\mathcal{E}(-3)$, so gilt, 4.4,

$$\omega_{ij}\eta_\mu = \sum_\nu a_{ij}^{\mu\nu}\xi_\nu \quad .$$

(β) Sind für $d \geq -1$ Basisvektoren $\xi_\mu^d \in H^1\mathcal{E}(d)$ ausgezeichnet, so gilt für die Koeffizienten $a_i^{\mu\nu} = {}^{-d-2}a_i^{\mu\nu}$ die Gleichung

$$z_i \xi_\mu^d = \sum_\nu a_i^{\mu\nu} \xi_\nu^{d+1} .$$

(γ) Die dualen Basiselemente von $H^2\mathcal{E}(-d-4) = H^1\mathcal{E}(d)^*$ definieren dann lediglich die "transponierten" Matrizen A_2^ε für $1 \leq \varepsilon \leq e$.

(4.7.2) Die Matrizen A_s^ε, $A_{1,1}^o$ kann man nach dieser Bemerkung nun auch direkt gewinnen als Darstellungen der folgenden Homomorphismen.

(α) Sei $W_{1,1}$ die 4×4-Matrix

$$\begin{pmatrix} -\omega_{12} & \omega_{o2} & -\omega_{o1} & 0 \\ -\omega_{13} & \omega_{o3} & 0 & -\omega_{o1} \\ -\omega_{23} & 0 & \omega_{o3} & -\omega_{o2} \\ 0 & -\omega_{23} & \omega_{13} & \omega_{12} \end{pmatrix}$$

der Basiselemente $\omega_{ij} \in \Gamma(\mathbb{P}_3, \Omega^1(2))$. Aufgrund der Operation ω_{ij}, 4.4, definiert diese Matrix den Homomorphismus

$$\alpha_{1,1}^o : 4H^2\mathcal{E}(-3) \xrightarrow[W_{1,1}]{} 4H^1\mathcal{E}(-1)$$

durch die "Multiplikation" $\xi \to \xi W_{1,1}$ der "Zeile" $\xi = (\xi_1, \dots, \xi_4)$ aus Vektoren von $H^2\mathcal{E}(-3)$ mit $W_{1,1}$. Nach der Basiswahl wird $\alpha_{1,1}^o$ durch $A_{1,1}^o$ dargestellt.

(β) Für $d \geq 1$ sei die Abbildung

$$\alpha_s^{-d} : \binom{4}{s+1} H^1\mathcal{E}(d-2) \xrightarrow[Z_s]{} \binom{4}{s} H^1\mathcal{E}(d-1)$$

analog definiert durch die Matrix Z_s, so daß α_s^{-d} induziert ist von dem Homomorphismus

$$\binom{4}{s+1}\mathcal{O}(d-2) \xrightarrow[Z_s]{} \binom{4}{s}\mathcal{O}(d-1) \ .$$

Nach Basiswahl wird α_s^{-d} repräsentiert von A_s^{-d}.

(γ) In analoger Weise interpretieren sich die Matrizen A_s^{ε} für $1 \leq \varepsilon \leq e$ bezüglich H^2.

Die Relationen (Q) folgen dann automatisch aus der Tatsache, daß die zusammengesetzten Homomorphismen

$$4H^2\mathcal{E}(-3) \xrightarrow[W_{1,1}]{} 4H^1\mathcal{E}(-1) \xrightarrow[Z_o]{} H^1\mathcal{E}$$

sowie

$$6H^1\mathcal{E}(d-2) \xrightarrow[Z_1]{} 4H^1\mathcal{E}(d-1) \xrightarrow[Z_o]{} H^1\mathcal{E}(d)$$

Null sind.

(4.8) Besondere Eigenschaften der Koeffizientensysteme für stabile 2-Bündel $\mathcal{E}$ über $\mathbb{P}_3$ mit $c_1 = 0$ und $H^1(\mathbb{P}_3, \mathcal{E}(-2)) = 0$.

Ist $\mathcal{E}$ ein solches Bündel, so gilt $\Gamma(\mathbb{P}_3, \mathcal{E}) = 0$ und $k = c_2 > 0$, vgl. [6]. Bei $c_1 = 0$ ist die Bedingung $\Gamma(\mathbb{P}_3, \mathcal{E}) = 0$ sogar gleichbedeutend mit der Stabilität. Wir setzen:

$M_I(0,k)$ = Menge der Isomorphieklassen der stabilen 2-Bündel über $\mathbb{P}_3$ mit $c_1 = 0$, $c_2 = k > 0$ und $H^1(\mathbb{P}_3, \mathcal{E}(-2)) = 0$

Die Menge $M_I(0,k)$ ist eine Teilmenge des groben Modulschemas $M(0,k)$ aller stabilen 2-Bündel über $\mathbb{P}_3$ mit $c_1 = 0$, $c_2 = k$, vgl. [16], die durch die Instantonenbedingung $H^1(\mathbb{P}_3, \mathcal{E}(-2)) = 0$ definiert ist.

(4.8.1) Ist $\mathcal{E} \in M_I(o,k)$, so folgt leicht unter Zuhilfenahme der Räume $H^i\Omega^1 \otimes \mathcal{E}(d)$, daß die Sequenzen

$$4H^2\mathcal{E}(-3) \xrightarrow[W_{1,1}]{} 4H^1\mathcal{E}(-1) \xrightarrow[Z_o]{} H^1\mathcal{E} \longrightarrow 0$$

und

$$6H^1\mathcal{E}(d-2) \xrightarrow[Z_1]{} 4H^1\mathcal{E}(d-1) \xrightarrow[Z_o]{} H^1\mathcal{E}(d) \to 0$$

für $d \geq 1$ sogar exakt sind. Insbesondere folgt dann sofort aus den Bemerkungen von 4.7.2, daß die Matrizen A_1^ε, $\varepsilon = -1,\ldots,-e$, durch $A_{1,1}^o$ (bis auf Basiswechsel) eindeutig dadurch bestimmt sind, daß die Spalten von A_o^{-1} den Lösungsraum von $A_{1,1}^o$ aufspannen und successive die Spalten von $A_o^{\varepsilon-1}$ den Lösungsraum von A_1^ε aufspannen. Damit ist ganz $Z^\ell(A)$ bereits durch $A_{1,1}^o$ bestimmt. Ferner erhält man aus den exakten Sequenzen sofort die Formeln

$$r_d = 4\ell_{d-1}^1 - \ell_d^1$$

für $d \geq 0$, denn $r_d = \text{rg } \alpha_1^{-d}$ für $d>0$ und $r_o = \text{rg } \alpha_{1,1}^o$.

(4.8.2) Ist $\mathcal{E} \in M_I(0,k)$, so folgt aus der Hirzebruch-Riemann-Roch Formel auch sofort, daß $\ell_{-1}^1 = \dim H^1\mathcal{E}(-1) = c_2 = k$ und $\ell_o^1 = \dim H^1\mathcal{E} = 2k-2$, so daß $r_o = 2k+2$.

<u>(4.8.3) Relationenkette von $A_{1,1}^o$</u>: Der Sachverhalt 4.8.1 führt nun direkt auf die Definition der "Relationenkette" einer gegebenen Matrix $A_{1,1}^o$ obigen Typs 4.6. Eine solche Matrix $A_{1,1}^o$ vom Rang $2k+2$ hat $q = 4k - (2k+2) = 2k-2$ unabhängige Lösungsspalten $\tilde{A}^\nu$ mit jeweils $4k$ Komponenten, $A_{1,1}^o \tilde{A}^\nu = 0$. Man kann dann $\tilde{A}^\nu$ in der Form

$$\tilde{A}^{\nu} = \begin{pmatrix} \tilde{A}_o^{1\nu} \\ \vdots \\ \tilde{A}_o^{k\nu} \end{pmatrix} \qquad \text{mit} \qquad \tilde{A}_o^{\mu\nu} = \begin{pmatrix} \tilde{a}_o^{\mu\nu} \\ \\ \tilde{a}_3^{\mu\nu} \end{pmatrix}$$

entsprechend der Aufteilung von $A_{1,1}^{o}$ schreiben. Dadurch erhält man für die Koeffizienten der Spalten $\tilde{A}^{\nu}$ eine Indizierung wie in 4.6, (2), und durch

$$A_1^{-1} = \begin{pmatrix} \tilde{A}_1^{11} & \cdots & \tilde{A}_1^{1q} \\ \vdots & & \vdots \\ A_1^{k1} & \cdots & \tilde{A}_1^{kq} \end{pmatrix}$$

ist dann die erste Matrix der Relationenkette bestimmt. In analoger Weise bestimme man nun eine Matrix A_1^{-2}, indem man von den Lösungen von A_1^{-1} anstelle der Lösungen von $A_{1,1}^{o}$ ausgeht. So fortfahrend erhalten wir eine Kette von Matrizen A_1^{-1}, $A_1^{-2}, \ldots, A_1^{-e}, \ldots$, die wir Relationenkette von $A_{1,1}^{o}$ nennen. Diese Matrizen sind bis auf Basistransformation eindeutig bestimmt.

(4.8.4) Zur natürlichen Zahl k sei $\mathfrak{M}_k$ die Menge der Matrizen

$$A_{1,1}^{o} = \begin{pmatrix} A_{1,1}^{11} & \cdots & A_{1,1}^{1k} \\ \vdots & & \vdots \\ A_{1,1}^{k1} & \cdots & A_{1,1}^{kk} \end{pmatrix}$$

wie oben mit den folgenden Bedingungen

(i) Für jedes Indexpaar μ,ν ist $A_{1,1}^{\mu\nu}$ eine Matrix vom Typ $A_{1,1}$ wie in 4.2 mit Koeffizienten $a_{ij}^{\mu\nu}$, $o \leq i < j \leq 3$.

(ii) $A^{\mu\nu}_{1,1} = A^{\nu\mu}_{1,1}$

(iii) rg $A^{o}_{1,1} = 2k + 2$

(iv) Die Relationenkette von $A^{o}_{1,1}$ gemäß 4.8.3 ist endlich.

Definition: Eine Matrix $A^{o}_{1,1} \in \mathfrak{M}_k$ heißt Instantonenmatrix.

(4.9) Die Surjektion $\mathfrak{M}_k \to M_I(0,k)$.
Ist $A^{o}_{1,1} \in \mathfrak{M}_k$ gegeben, so erhält man die Relationenkette $A_1^{-1},\ldots,A_1^{-e}$ mit den Rangzahlen $r_d = \text{rg } A_1^{-d}$ für $d = 1,\ldots,e$, sowie $r_o = 2k + 2$. Man setze dann $\ell^1_{-1} = k$, $\ell^1_o = 2k - 2,\ldots,$ $\ell^1_d = 4\ell^1_{d-1} - r_d$, für $d<e$. (Es ist $r_e = 4\ell^1_{e-1}$, da die Relationenkette bei A_1^{-e} abbricht.) Dann ist durch diese Daten ein Bündel $\mathcal{F}^{\ell}(A)$ definiert, das wie oben aufspaltet in

$$\mathcal{F}^{\ell}(A) = \mathcal{E}(A) \oplus \mathcal{L}.$$

Wegen der speziellen Situation $\ell^1_d = 4\ell^1_{d-1} - r_d$ folgt nun leicht, daß rg $\mathcal{E}(A) = 2$, $c_1\mathcal{E}(A) = 0$, $c_2\mathcal{E}(A) = k$, $\Gamma\mathcal{E}(A) = 0$, $H^1\mathcal{E}(A)(d) = 0$ für $d\leq -2$, so daß $\mathcal{E}(A)$ ein Element in $M_I(0,k)$ definiert. Da $\mathcal{E}(A)$ bis auf Isomorphie eindeutig durch $A^{o}_{1,1}$ bestimmt ist, ist die Abbildung

$$\mathfrak{M}_k \longrightarrow M_I(0,k)$$

wohldefiniert. Aus dem Darstellungssatz 4.7 und aus 4.8.1 folgt, daß diese Abbildung surjektiv ist.

(4.10) Auf dem Matrizenraum $\mathfrak{M}_k$ operiert die Gruppe $GL(k,\mathbb{C})$ wie folgt. Ist $P = (p_{\mu\nu}) \in GL(k,\mathbb{C})$, so sei für $A^{o}_{1,1} \in \mathfrak{M}_k$ die Matrix $B^{o}_{1,1}$ definiert durch

$$B^{\mu\nu}_{1,1} = \sum_{\lambda,\kappa} p_{\mu\lambda} A^{\lambda\kappa}_{1,1} p_{\nu\kappa}$$

oder kurz $B^o = PA^oP^t$. Es ist $B^o_{1,1} \in \mathfrak{M}_k$ und wir schreiben $A^o \sim B^o$. Wenn dies der Fall ist, so folgt $\mathcal{Z}^\ell(A) \cong \mathcal{Z}^\ell(B)$ und damit auch $\mathcal{E}(A) \cong \mathcal{E}(B)$. Die Umkehrung folgt nach 4.7.1 auch, wobei man P als Basistransformation in $H^1\mathcal{E}(-1)$ zu interpretieren hat. Mit diesem Resultat folgt schließlich, daß $\mathfrak{M}_k/GL(k,\mathbb{C}) \cong M_I(0,k)$.

Bemerkung: Die hier gegebene Parametrisierung von $M_I(0,k)$ entspricht derjenigen, die durch Monaden gegeben ist, [7]. Die wesentliche Schwierigkeit bei der Untersuchung von $M_I(0,k)$ liegt dabei in der Charakterisierung der Eigenschaft (iv) von $\mathfrak{M}_k$. Diese Bedingung entspricht der Monadenbedingung in [7]. Für den einfachsten Fall k = 1, 4.12, sind in § 6 alle Instantonenfelder berechnet. Der Modulraum $M_I(0,2)$ ist mit anderen Methoden in [11] behandelt.

(4.11) Koordinatenwechsel: Zur Definition der $Z^\ell(A)$ haben wir die homogenen Koordinaten $z_o,\dots,z_3$ des $\mathbb{P}_3$ ausgezeichnet. Baut man für dasselbe Bündel eine Matrix $W^\ell(B)$ vermöge der Koordinaten $w_o,\dots,w_3$ und der Operatoren w_i, ω_{ij} wie in 4.7.1 auf, so hat man die Koeffizienten $a^{\mu\nu}_{ij}$, $a^{\mu\nu}_i$ entsprechend den z_i, ω_{ij} zu transformieren. Sei etwa $W_o = CZ_o$ und $C : \mathbb{C}^4 \to \mathbb{C}^4$ die Transformationsmatrix der homogenen Koordinaten, und seien $\tilde{\omega}_{ij}$ die durch w_i gegebenen Basiselemente von $\Gamma\Omega^1(2)$.
Sei $C^{(s)} : \wedge^s\mathbb{C}^4 \to \wedge^s\mathbb{C}^4$ die induzierte Matrix der s-reihigen Unterdeterminanten von C. Dann folgt für die Spalten $\tilde{\omega} = (\tilde{\omega}_{ij})$

und $\omega = (\omega_{ij})$ die Beziehung $\tilde{\omega} = C^{(2)}\omega$. Für die Matrizen $B^{\mu\nu}_{t,s}$ des neuen Koeffizientensystems gilt dann:

$$B^{\mu\nu}_{t,o} = C^{(t+1)}A^{\mu\nu}_{t,o}$$

$$B^{\mu\nu}_{t,s} = C^{(t+s+1)}A^{\mu\nu}_{t,s}C^{(s)-1} .$$

(4.12) Der Fall k = 1. Hier identifiziert sich $\mathfrak{M}_1$ mit der Menge der Matrizen

$$A_{1,1} = \begin{pmatrix} -a_{12} & a_{o2} & -a_{o1} & 0 \\ -a_{13} & a_{o3} & 0 & -a_{o1} \\ -a_{23} & 0 & a_{o3} & -a_{o2} \\ 0 & -a_{23} & a_{13} & -a_{12} \end{pmatrix}$$

mit rg $A_{1,1} = 4$, d.h. $a_{o1}a_{23} - a_{o2}a_{13} + a_{o3}a_{12} \neq 0$, und $M_I(0,1)$ ist dann nichts weiter als die Mannigfaltigkeit $\mathbb{P}_5 \setminus Q$, vgl. § 2, die der Quotient von $\mathfrak{M}_1$ bezüglich der Gruppe $\mathbb{C}^*$ ist. Entspricht $\mathcal{E} \in M_I(0,1)$ dem Punkt $(a_{o1} : a_{o2} : \ldots : a_{23}) \in \mathbb{P}_5 \setminus Q$, so hat $\mathcal{E}$ eine Darstellung

$$\mathcal{E} \oplus 4\mathcal{O} \cong \mathcal{Z}(A)$$

mit

$$Z(A) = \begin{pmatrix} z_1 & 0 \\ A_{1,1} & z_2 \end{pmatrix} .$$

(4.13) Für Bündel $\mathcal{E}$ des Typs 4.12 kann man nun leicht die Beschränkungen $\mathcal{E}|L_p$ auf Geraden $L_p \subset \mathbb{P}_3$ bestimmen. Zu diesem Zweck bezeichne $P_{1,s}$ die aus den Plückerkoordinaten p_{ij} von $p \in Q$ gebildete Matrix vom Typ $A_{1,s}$. Es gilt:

(1) $\mathcal{E}|L_p = 2\,\mathcal{O}_{L_p}$ genau dann, wenn

$\delta := P_{1,2}A_{1,o} = p_{o1}a_{23} - p_{o2}a_{13} + p_{o3}a_{12} + p_{12}a_{o3} - p_{13}a_{o2} + p_{23}a_{o1} \neq 0$

(2) Wenn $\delta = 0$, so ist $\mathcal{E}|L_p = \mathcal{O}_{L_p}(-1) \oplus \mathcal{O}_{L_p}(1)$.

Beweisskizze: Ist die Gerade L_p durch $z_2 = z_3 = 0$ gegeben, so kann man zeigen, daß $\mathcal{F}(A) \mid L_p \cong 6\,\mathcal{O}_{L_p}$ ist genau dann, wenn $a_{23} \neq 0$,[23].Ist $x \neq y \in \mathbb{P}_3$ beliebig, so existiert eine Koordinatentransformation C mit $\tilde{x} = Cx = (1,0,0,0)$ und $\tilde{y} = Cy = (0,1,0,0)$. Dann gilt für die Plückerkoordinaten $\tilde{p}_{ij}$ von $\tilde{x},\tilde{y}$: $\tilde{p}_{o1} = 1$ und $\tilde{p}_{ij} = 0$ sonst. Ist $\tilde{A}_{1,o}$ die Matrix von $\mathcal{E}$ in den neuen Koordinaten, so ist $\tilde{A}_{1,o} = C^{(2)}A_{1,o}$ sowie $\tilde{P}_{1,2} = C^{(4)}P_{1,2}C^{(2)-1}$, woraus folgt, daß

$\tilde{a}_{23} = \tilde{P}_{1,2}\tilde{A}_{1,o} = (\det C)\, P_{1,2}A_{1,o}$. Daher ist $\mathcal{F}(A)|L_p = 6\mathcal{O}_{L_p}$ genau dann, wenn $\tilde{a}_{23} \neq 0$, d.h. $\delta \neq 0$. Wenn $\delta = 0$ oder $\tilde{a}_{23} = 0$, so folgt $\mathcal{F}(A)|L_p = 4\mathcal{O}_{L_p} \oplus \mathcal{O}_{L_p}(-1) \oplus \mathcal{O}_{L_p}(1)$, [23].

Bemerkung: Die Bedingung $\delta = 0$ besagt gerade, daß $p \in Q \cap H_a$, wo H_a die Polare des Punktes $a \in \mathbb{P}_5$ bezüglich der Quadrik Q ist.

§ 5 Symplektische Involutionen.

Sei wie oben $\sigma : \mathbb{P}_3 \to \mathbb{P}_3$ die Involution $(z_o\ z_1\ z_2\ z_3) \longmapsto (-\bar{z}_1\ \bar{z}_o\ -\bar{z}_3\ \bar{z}_2)$. Setzt man $\sigma(o) = 1$, $\sigma(1) = 0$, $\sigma(2) = 3$, $\sigma(3) = 2$ und $U_i = \{z_i \neq 0\}$, so gilt $\sigma^{-1}U_i = U_{\sigma(i)}$ und σ wird beschrieben durch $z \longrightarrow w$ mit $w_{\sigma(i)} = (-1)^i\ \bar{z}_i$ oder $w_i = -(-1)^i\ \bar{z}_{\sigma(i)}$.

Ist E ein holomorphes Vektorbündel mit Cozyklus g_{ij} bzgl. (U_i), so wird $\sigma^*\bar{E}$ durch $g_{ij}^\sigma := \bar{g}_{\sigma(i)\sigma(j)} \circ \sigma$ beschrieben und ist wieder ein holomorphes Bündel. Ist $h : E \to F$ ein holomorpher Homomorphismus mit den lokalen Homomorphismen $h_i : \mathcal{O}^r \to \mathcal{O}^s$ über U_i, so wird $\sigma^*\bar{h} : \sigma^*\bar{E} \to \sigma^*\bar{F}$ beschrieben durch $h_i^\sigma = \bar{h}_{\sigma(i)} \circ \sigma$. Analog kann man den kanonischen antilinearen Isomorphismus

$$\Gamma(\mathbb{P}_3, E) \longrightarrow \Gamma(\mathbb{P}_3, \sigma^*\bar{E})$$

beschreiben durch $(s_i) \longleftrightarrow (s_i^\sigma)$, wo $s_i \in \Gamma(U_i, \mathcal{O}^r)$ mit $s_i = g_{ij}s_j$ und $s_i^\sigma = \bar{s}_{\sigma(i)} \circ \sigma$.

(5.1) Unter einer symplektischen Involution auf einem holomorphen Vektorbündel E verstehen wir einen holomorphen Homomorphismus $\sigma : E \to \sigma^*\bar{E}$ mit $\sigma^2 = -1$, d.h. $\sigma^*(\bar{\sigma}) \circ \sigma = -\mathrm{id}$, wobei man beachte, daß $\sigma^*\overline{\sigma^*\bar{E}} = E$. Ist $\sigma^2 = +1$, so heißt σ eine reelle Involution.

Jeder solche Homomorphismus induziert eine antilineare Abbildung

$$\sigma : \Gamma(\mathbb{P}_3, E) \longrightarrow \Gamma(\mathbb{P}_3, E)$$

mit $\sigma^2 = -1$ vermöge $(s_i) \to (\tilde{s}_i)$, wobei

$$\tilde{s}_i = (\overline{\sigma_{\sigma(i)} s_{\sigma(i)}}) \circ \sigma$$

und wobei $\sigma_i : \mathcal{O}^r \to \mathcal{O}^r$ die lokale holomorphe Matrix über U_i von $\sigma : E \to \sigma^* \bar{E}$ ist, $\sigma_i g_{ij} = g_{ij}^{\sigma} \sigma_j$. Dies folgt durch Komposition der Abbildungen $\Gamma(\mathbb{P}_3,E) \xrightarrow[\Gamma\sigma]{} \Gamma(\mathbb{P}_3,\sigma^* \bar{E}) \xleftarrow[=]{} \Gamma(\mathbb{P}_3,E)$.

(5.2) Ist E ein Instantonenbündel, 3.5, welches nicht trivial ist, so gilt $c_1 = 0$, $c_2 > 0$ und $\Gamma(\mathbb{P}_3,E) = 0$, so daß E stabil ist und die Endomorphismen von E nur aus den Homothetien mit Konstanten bestehen.

Beweis: Da $E|L_p \cong 2\,\mathcal{O}|L_p$ für jede reelle Gerade L_p, ist $c_1 = 0$ und $\Gamma(L_p, E(-1)|L_p) = 0$. Daraus folgt $\Gamma(\mathbb{P}_3, E(-1)) = 0$. Wäre $0 \neq s \in \Gamma(\mathbb{P}_3, E)$, so folgte leicht aus $\Gamma(\mathbb{P}_3, E(-1)) = 0$, daß entweder $\{s = 0\}$ leer ist oder die Codimension 2 hat. Bezeichnet F die Quotientengarbe der durch s gegebenen Einbettung $\mathcal{O} \hookrightarrow E$, so folgt im Falle $\{s = 0\} = \emptyset$, daß F ein Geradenbündel und wegen $c_1F = 0$ sogar $F \cong \mathcal{O}$ ist. Dann wäre aber $E \cong 2\mathcal{O}$, was ausgeschlossen wurde. Dann hätte $\{s = 0\}$ die Codimension 2 und wegen $c_1E = 0$ müßte F eine Idealgarbe sein mit $\Gamma(\mathbb{P}_3,F) = 0$, [9] [10]. Das implizierte $\Gamma(\mathbb{P}_3,\mathcal{O}) \cong \Gamma(\mathbb{P}_3,E)$. Wegen der symplektischen Involution σ aber müßte $\Gamma(\mathbb{P}_3,E)$ eine gerade Dimension haben. Die Stabilität von E sowie $c_2 > 0$ folgt nun aus [6].

(5.3) Nach einem <u>Satz von Atiyah-Hitchin-Drinfeld-Manin</u> [3] gilt für ein Instantonenbündel darüberhinaus der wichtige Verschwindungssatz $H^1(\mathbb{P}_3,E(d)) = 0$, $d \leq -2$.

<u>(5.4) Lemma:</u> Ist E ein stabiles 2-Bündel über $\mathbb{P}_3$ mit einer symplektischen Involution $\sigma : E \to \sigma^* \bar{E}$, so ist σ bis auf eine Konstante c, $|c| = 1$, eindeutig bestimmt.

Beweis: Sind σ, τ zwei solche Involutionen, so ist $\sigma^{-1}\tau : E \to E$ ein holomorpher Automorphismus, der wegen der Stabilität von E eine Konstante c ist, d.h. $\tau = c\sigma$. Aus $\tau^2 = -1$ und $\sigma^2 = -1$ folgt dann $|c| = 1$.

(5.5) Ist $E = \mathcal{O}(d)$, $d \in \mathbb{Z}$, mit dem Cozyklus $g_{ij} = (\frac{z_j}{z_i})^d$, so wird durch $\sigma_i = (-1)^{id}$ eine Involution $\sigma : \mathcal{O}(d) \to \sigma^*\overline{\mathcal{O}(d)}$ definiert, deren lokale Matrix über U_i gerade σ_i ist. Es ist $\sigma^2 = (-1)^d$. Das Bündel $\mathcal{O}(d)$ hat also eine reelle (symplektische) Involution, wenn d gerade (ungerade) ist. Die auf $\Gamma\,\mathcal{O}(d)$ induzierte antilineare Involution für $d \geq 0$ berechnet sich nach 5.1 zu $a \longrightarrow (-1)^d\, \bar{a} \circ \sigma$, wo a ein homogenes Polynom vom Grad d ist.

Ist insbesondere $a = a_0z_0 + \ldots + a_3z_3$ linear, so hat $b = -\bar{a} \circ \sigma = b_0z_0 + \ldots + b_3z_3$ die Koeffizienten $b_0 = -\bar{a}_1$, $b_1 = \bar{a}_0$, $b_2 = -\bar{a}_3$, $b_3 = \bar{a}_2$.

(5.6) Bemerkung: Sind $\sigma : E \to \sigma^*\bar{E}$ und $\tau : F \to \sigma^*\bar{F}$ Involutionen holomorpher Vektorbündel, so erhält man durch $\sigma \otimes \tau : E \otimes F \longrightarrow \sigma^*\bar{E} \otimes \sigma^*\bar{F} = \sigma^*(\overline{E \otimes F})$ eine Involution mit $(\sigma \otimes \tau)^2 = \sigma^2 \otimes \tau^2$. Insbesondere hat E(d) eine reelle (symplektische) Involution, wenn E eine symplektische Involution besitzt und d ungerade (gerade) ist.

(5.7) Der Koszul-Komplex im $\mathbb{P}_3$ besitzt eine kanonische reelle Involution wie folgt

$$\begin{array}{ccccccccccccc}
0 \longrightarrow & \mathcal{O}(-4) & \xrightarrow{Z_3} & 4\mathcal{O}(-3) & \xrightarrow{Z_2} & 6\mathcal{O}(-2) & \xrightarrow{Z_1} & 4\mathcal{O}(-1) & \xrightarrow{Z_0} & \mathcal{O} & \to 0 \\
& \downarrow{\scriptstyle \Sigma_4} & & \downarrow{\scriptstyle \Sigma_3} & & \downarrow{\scriptstyle \Sigma_2} & & \downarrow{\scriptstyle \Sigma_1} & & \downarrow{\scriptstyle \Sigma_0} & \\
0 \longrightarrow & \sigma^*\overline{\mathcal{O}(-4)} & \xrightarrow[Z_3^\sigma]{} & 4\sigma^*\overline{\mathcal{O}(-3)} & \xrightarrow[Z_2^\sigma]{} & 6\sigma^*\overline{\mathcal{O}(-2)} & \xrightarrow[Z_1^\sigma]{} & 4\sigma^*\overline{\mathcal{O}(-1)} & \xrightarrow[Z_0^\sigma]{} & \sigma^*\overline{\mathcal{O}} & \to 0\,,
\end{array}$$

wobei die lokalen Matrizen $\Sigma_{\nu,i}$ von Σ_ν über U_i so gegeben sind:

$$\Sigma_{o,i} = 1\,, \quad \Sigma_{1,i} = (-1)^i \left(\begin{array}{cc|cc} 0 & 1 & 0 & 0 \\ -1 & 0 & 0 & 0 \\ \hline 0 & 0 & 0 & 1 \\ 0 & 0 & -1 & 0 \end{array}\right)$$

$$\Sigma_{2,i} = \left(\begin{array}{c|cccc|c} 1 & 0 & 0 & 0 & 0 & 0 \\ \hline 0 & 0 & 0 & 0 & 1 & 0 \\ 0 & 0 & 0 & -1 & 0 & 0 \\ 0 & 0 & -1 & 0 & 0 & 0 \\ 0 & 1 & 0 & 0 & 0 & 0 \\ \hline 0 & 0 & 0 & 0 & 0 & 1 \end{array}\right)$$

$$\Sigma_{3,i} = (-1)^i \left(\begin{array}{cc|cc} 0 & 1 & 0 & 0 \\ -1 & 0 & 0 & 0 \\ \hline 0 & 0 & 0 & 1 \\ 0 & 0 & -1 & 0 \end{array}\right) \qquad \Sigma_{4,i} = 1\,.$$

Der Leser kann verifizieren, daß mit diesen Matrizen das obige Diagramm kommutiert und daß $\Sigma_\nu^2 = \mathrm{id}$. Dabei ist, vgl. Vorbemerkung zu diesem Paragraphen, $Z_\nu^\sigma = \bar{Z}_\nu \circ \sigma$. Durch die Homomorphismen Σ_p des Koszul-Komplexes werden reelle Involutionen $\Omega^p \longrightarrow \sigma^*\overline{\Omega^p}$ induziert, sowie Involutionen $\Omega^p(d) \to \sigma^*\overline{\Omega^p(d)}$ für alle $d \in \mathbb{Z}$. Insbesondere folgt

(5.7.1) Die kanonische Basis $\omega_{o1} = (-z_1, z_o, 0, 0), \ldots$ von $\Gamma\Omega^1(2)$ wird vermöge des antilinearen Isomorphismus $\Sigma_1(2)$ wie folgt transformiert:

$$\omega_{o1} \to \omega_{o1}, \ \omega_{o2} \to \omega_{13}, \ \omega_{12} \to -\omega_{o3},$$

$$\omega_{13} \to \omega_{o2}, \ \omega_{o3} \to -\omega_{12}, \ \omega_{23} \to \omega_{23} .$$

(5.7.2) Die antilineare Involution $\Sigma_2(3): \Gamma 6\mathcal{O}(1) \to \Gamma 6\mathcal{O}(1)$ wird durch die Zuordnung $(f_{23}, -f_{13}, f_{o3}, f_{12}, -f_{o2}, f_{o1}) \to$ $(g_{23}, -g_{13}, g_{o3}, g_{12}, -g_{o2}, g_{o1})$ mit

$$g_{23} = \overline{f}_{23}\circ\sigma, \ g_{13} = \overline{f}_{o2}\circ\sigma, \ g_{o3} = -\overline{f}_{12}\circ\sigma$$

$$g_{o2} = \overline{f}_{13}\circ\sigma, \ g_{12} = -\overline{f}_{o3}\circ\sigma, \ g_{o1} = \overline{f}_{o1}\circ\sigma$$

gegeben, wobei f_{ij}, g_{ij} Linearformen sind.

(5.8) σ-ausgezeichnete Basen der Cohomologiegruppen:
Ist V ein endlich-dimensionaler $\mathbb{C}$-Vektorraum und ist σ: V $\to$ V eine antilineare Involution mit $\sigma^2 = 1$, so besitzt V eine Basis, die unter σ invariant ist. Ist dagegen $\sigma^2 = -1$, so hat V eine gerade Dimension 2n und eine Basis $\xi_1, \xi_2, \ldots, \xi_{2n-1}, \xi_{2n}$ mit $\sigma\xi_{2\nu-1} = \xi_{2\nu}$ und $\sigma\xi_{2\nu} = -\xi_{2\nu-1}$. Solche Basen sollen σ-ausgezeichnet heißen. Ist nun E ein holomorphes Vektorbündel über $\mathbb{P}_3$ mit einer symplektischen Involution $\sigma : E \to \sigma^*\overline{E}$, so werden dadurch (wie oben für $\Gamma(\mathbb{P}_3, E)$) antilineare Involutionen $\sigma : H^i(\mathbb{P}_3, E(d)) \to H^i(\mathbb{P}_3, E(d))$ induziert. Diese Involutionen sind auch verträglich mit der Serre-Dualitäts-Paarung, die hier auf $E \times E \to \Lambda^2 E \cong \mathcal{O}$ beruht, 3.5. Ist insbesondere E ein Instantonenbündel und wählt man in $H^1(\mathbb{P}_3, E(-1)), \ldots, H^1(\mathbb{P}_3, E(e-1))$ σ-ausgezeichnete Basen, so folgt leicht, daß auch die Serre-dualen Basen von $H^2(\mathbb{P}_3, E(-3)), \ldots, H^2(P_3, E(-e-3))$ σ-ausgezeichnet sind.

Betrachtet man dann die Koeffizienten $a_{ij}^{\mu\nu}$, ${}^{\varepsilon}a_{i}^{\mu\nu}$ die zu E und solchen Basen nach 4.7.1 gegeben sind, so ergibt sich im einzelnen:

(5.8.1) Die Koeffizienten $a_{ij} = a_{ij}^{\mu\nu}$ der Teilmatrix $A_{1,1}^{o}$ erfüllen

$$a_{o1} = \bar{a}_{o1},\ a_{o2} = \bar{a}_{13},\ a_{12} = -\bar{a}_{o3},\ a_{23} = \bar{a}_{23} .$$

Diese Gleichungen kann man auch durch $A_{1,o}^{\mu\nu} = \sigma A_{1,o}^{\mu\nu}$ ausdrücken, wo σ die reelle Struktur des $\mathbb{P}_5$ oder $\mathbb{C}^6$ in den Koordinaten p_{ij} ist, § 2.

(5.8.2) Für die Koeffizienten eines Blocks A_1^{ε} mit $-e \leq \varepsilon \leq -1$ gilt:

(i) Ist ε ungerade und sind $a_i = {}^{\varepsilon}a_i^{\mu,2\nu}$, $b_i = {}^{\varepsilon}a_i^{\mu,2\nu-1}$, so gilt

$$a_o = -\bar{b}_1,\ a_1 = \bar{b}_o,\ a_2 = -\bar{b}_3,\ a_3 = \bar{b}_2 .$$

Diese Gleichungen kann man analog durch $\sigma\,{}^{\varepsilon}A_o^{\mu,2\nu-1} = {}^{\varepsilon}A_o^{\mu,2\nu}$ zusammenfassen.

(ii) Ist ε gerade und sind $a_i = {}^{\varepsilon}a_i^{2\mu,\nu}$, $b_i = {}^{\varepsilon}a_i^{2\mu-1,\nu}$, so gilt entsprechend

$$a_o = -\bar{b}_1,\ a_1 = \bar{b}_o,\ a_2 = -\bar{b}_3,\ a_3 = \bar{b}_2$$

oder $\sigma\,{}^{\varepsilon}A_o^{2\mu-1,\nu} = {}^{\varepsilon}A_o^{2\mu,\nu}$.

(5.8.3) Durch 5.8.2 sind gleichzeitig die Bedingungen für die Matrizen A^{ε} mit $1 \leq \varepsilon \leq e$ gegeben, die sich aus $A^{-\varepsilon}$ durch Transposition ergeben, 4.6.

Beweis: Die Operationen der $\omega_{ij} \in \Gamma\Omega^1(2)$ auf $H^2E(-3)$ sind kanonisch mit den symplektischen (reellen) Involutionen der einzelnen Gruppen verträglich. Ist nun wie in 4.7.1

$$\omega_{ij}\eta_\mu = \sum_\nu a_{ij}^{\mu\nu}\xi_\nu ,$$

und sind die Basen reell, so folgt

$$\sigma(\omega_{ij})\eta_\mu = \sum_\sigma \bar{a}_{ij}^{\mu\nu}\xi_\nu .$$

Ist nun $\sigma(\omega_{ij}) = \pm\omega_{k\ell}$, vgl. 5.7.1, so folgt

$$\pm a_{k\ell}^{\mu\nu} = \bar{a}_{ij}^{\mu\nu}$$

woraus 5.8.1 folgt. Analog ergibt sich 5.8.2. Denn die Matrix A_1^ε wird durch die Abbildung $Z_1 : 6H^1E(-\varepsilon-2) \rightarrow 4H^1E(-\varepsilon-1)$ gegeben, und ist ε ungerade, so hat $H^1E(-\varepsilon-1)$ eine symplektische Struktur und eine dazu ausgezeichnete Basis $\ldots\eta_{2\nu-1}, \eta_{2\nu}\ldots$. Ist $\ldots\xi_\mu\ldots$ die reelle Basis von $H^1E(-\varepsilon-2)$, so gilt

$$z_i\xi_\mu = \sum_\nu (a_i^{\mu,2\nu-1}\eta_{2\nu-1} + a_i^{\mu,2\nu}\eta_{2\nu}) .$$

Dann folgt wie eben

$$\sigma(z_i)\xi_\mu = \sum_\nu (\bar{a}_i^{\mu,2\nu-1}\eta_{2\nu} - \bar{a}_i^{\mu,2\nu}\eta_{2\nu-1}) .$$

Ist dann $\sigma(z_i) = \pm z_k$ (σ ist hier die antilineare Abbildung von $\Gamma\mathcal{O}(1)$ in sich), so folgt daraus $\bar{a}_i^{\mu,2\nu-1} = \pm a_k^{\mu,2\nu}$, d.h. 5.8.2 .

(5.9) Insgesamt erhält man so eine antilineare Involution Σ für die obigen Koeffizientensysteme A, indem man die Bedingungen 5.8.1, 5.8.2 zu $\Sigma A = A$ zusammenfaßt. Insbesondere wird dadurch eine reelle Struktur $\Sigma: \mathfrak{M}_k \rightarrow \mathfrak{M}_k$ induziert, so daß eine Matrix $A_{1,1}^O$ reell ist genau dann, wenn die Koeffizienten $a_{ij}^{\mu\nu}$ die Bedingung 5.8.1 erfüllen. Ist $A_{1,1}^O \in \mathfrak{M}_k$ mit $\Sigma A_{1,1}^O = A_{1,1}^O$ gegeben, so kann man auch die Matrizen A_1^ε der Relationenkette von $A_{1,1}^O$ so wählen, daß 5.8.2 erfüllt ist. Dann definiert $A_{1,1}^O$ auch umgekehrt ein Bündel $E \in M_I(O,k)$ mit einer symplektischen Involution $\sigma : E \to \sigma^*\overline{E}$. Denn:

(5.10) Erfüllt das Koeffizientensystem A umgekehrt die Bedingung $\Sigma A = A$, d.h. 5.8.1, 5.8.2, und ist $\mathcal{E}$ durch die Zerlegung $\mathcal{E} \oplus \mathcal{L} \cong \mathcal{Z}^\ell(A)$ definiert, so wird eine symplektische Involution $\sigma : \mathcal{E} \to \sigma^*\overline{\mathcal{E}}$ wie folgt induziert. Durch direkte Summen der Involutionen Σ_p des Koszul-Komplexes erhält man wegen 5.8.1, 5.8.2 das kommutative Diagramm

$$\begin{array}{ccc} \mathcal{L}_1 & \xrightarrow{\;z^\ell(A)\;} & \mathcal{L}_o \\ \Big\downarrow \sigma^{(1)} & & \Big\downarrow \sigma^{(o)} \\ \sigma^*\overline{\mathcal{L}}_1 & \xrightarrow[\;z^\ell(A)^\sigma\;]{} & \sigma^*\overline{\mathcal{L}}_o \end{array}$$

mit symplektischen Involutionen $\sigma^{(o)}, \sigma^{(1)}$, wo $z^\ell(A)^\sigma = \overline{z^\ell(A)} \circ \sigma$, vgl. Vorbemerkung zu § 5. Daraus resultiert eine symplektische Involution

$$\sigma : \mathcal{Z}^\ell(A) \to \sigma^*\overline{\mathcal{Z}^\ell(A)}$$

für das Bild des Homomorphismus $z^\ell(A)$. Diese Involution induziert die gewünschte Involution auf $\mathcal{E}$ und liefert über die Operatoren ω_{ij}, z_i die ursprünglichen Bedingungen zurück.

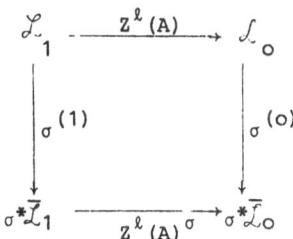

Aus dem Darstellungssatz 4.7 und aus 5.8, 5.10 ergibt sich nun eine Klassifizierung der Instantonenbündel (bzw. der selbstdualen euklidischen SU(2)-Yang-Mills Felder), indem man die Koeffizientensysteme $A^{o}_{1,1} \in \mathcal{M}_k$ betrachtet, die $\Sigma A^{o}_{1,1} = A^{o}_{1,1}$ erfüllen und deren zugeordnete Bündel E auf reellen Geraden trivial sind, vgl. 4.13. Im folgenden wird diese Klassifikation benutzt, um im Fall $c_2 = 1$ die zugehörigen Yang-Mills Felder abzuleiten, die stets auch bei beliebigem $c_2 > 0$ rationale Ausdrücke in den Koeffizienten des Systems A werden.

§ 6 Herleitung der t'Hooft-Lösung für k=1.

Ist $\mathcal{E}$ ein Instantonenbündel auf $\mathbb{P}_3$ mit $c_2 = 1$, so ist $\mathcal{E}$ stabil und hat nach 4.12 eine Darstellung

$$\mathcal{E} \oplus 4\mathcal{O} \cong \mathcal{Z}(A) \quad \text{mit } Z(A) = \begin{pmatrix} Z_1 & 0 \\ A_{1,1} & Z_2 \end{pmatrix} .$$

Für die Koeffizienten $a_{01},\ldots,a_{23}$ von $A_{1,1}$ bzw. $A_{1,0}$ gelten dann die Bedingungen

(i) $d = a_{01}a_{23} - a_{02}a_{13} + a_{03}a_{12} \neq 0$

(ii) Für jedes reelle $p \in S^4$ ist

$$\delta(p) = p_{01}a_{23} - p_{02}a_{13} + p_{03}a_{12} + p_{12}a_{03} - p_{13}a_{02} + p_{23}a_{01} \neq 0$$

(iii) $a_{01} = \bar{a}_{01},\ a_{02} = \bar{a}_{13},\ a_{12} = -\bar{a}_{03},\ a_{23} = \bar{a}_{23}$.

Die Bedingung (i) folgt aus dem Darstellungssatz wegen $\operatorname{rg} \mathcal{E} = 2$, die Bedingung (ii) folgt aus 4.13, da $E|L_p$ trivial ist für reelle p. Die Bedingung (iii) ergibt sich aus der symplektischen Struktur von $\mathcal{E}$ über 5.8.1. Aus (ii) und (iii) folgen sogar

(i)* $d = a_{01}a_{23} - a_{02}a_{13} + a_{03}a_{12} > 0$

(ii)* $\delta(p) > 0$ für alle $p \in S^4$.

Denn benutzt man die Koordinaten $x_0,\ldots,x_5$ des $\mathbb{P}_5$ nach 2.1 für p und entsprechende Koordinaten $a_0,\ldots,a_5$ für $a = (a_{01},\ldots,a_{23})$, so ist

$$2\delta(p) = x_0a_0 - x_1a_1 - \ldots - x_5a_5$$

mit reellen a_i und x_i und mit

$$x_o^2 = x_1^2 + \ldots + x_5^2 .$$

Es folgt dann leicht, daß die Bedingung $\delta(p) \neq 0$ für alle diese x genau dann erfüllt ist, wenn

$$\sum_{\nu=1}^{5} \left(\frac{a_\nu}{a_o}\right)^2 < 1 ,$$

was gleichbedeutend mit $(i)^*$ ist. Es ist dann

$$\sum_{\nu=1}^{5} \frac{x_\nu}{x_o} \frac{a_\nu}{a_o} < 1$$

für alle $\frac{x}{x_o} \in S^4$, und diese Bedingung ist gleichbedeutend mit $(ii)^*$.

(6.1) Es gibt eine exakte Sequenz

$$0 \dashrightarrow \mathcal{E} \longrightarrow 6\,\mathcal{O}(1) \xrightarrow[(A_{1,o}Z_1)]{} \mathcal{O}(1) \oplus 4\,\mathcal{O}(2)$$

durch die auf $\mathcal{E}$ die symplektische Involution induziert wird.

Beweis: Sei zunächst $\mathcal{E}'$ die Kerngarbe des Homomorphismus $(A_{1,o}, Z_1)$. Dann erhält man das kommutative Diagramm

$$\begin{array}{ccccccc}
0 & \dashrightarrow & \mathfrak{Z}(A) & \longrightarrow & 4\,\mathcal{O} \oplus 6\,\mathcal{O}(1) & \xrightarrow{F} & \mathcal{O}(1) \oplus 4\,\mathcal{O}(2) \\
 & & \downarrow \mu & & \downarrow M & & \| \\
0 & \dashrightarrow & 4\,\mathcal{O} \oplus \mathcal{E}' & \dashrightarrow & 4\,\mathcal{O} \oplus 6\,\mathcal{O}(1) & \xrightarrow[F']{} & \mathcal{O}(1) \oplus 4\,\mathcal{O}(2)
\end{array} ,$$

wobei für die Matrizen M, F, F' gilt MF' = F mit

$$F = \begin{pmatrix} Z_o & 0 \\ A_{1,o} & Z_1 \end{pmatrix} \qquad F' = \begin{pmatrix} 0 & 0 \\ A_{1,o} & Z_1 \end{pmatrix} \qquad M = \begin{pmatrix} I & -A_{1,1}^{-1} Z_2 \\ 0 & I \end{pmatrix} .$$

Dabei ist $Z_o = -A_{1,1}^{-1} Z_2 A_{1,o}$ wegen der Relation 4.2.1. Die invertierbare Matrix M induziert nun den Isomorphismus μ , so daß $4\mathcal{O} \oplus \mathcal{E} \cong \mathcal{Z}(A) \cong 4\mathcal{O} \oplus \mathcal{E}'$, woraus folgt, daß $\mathcal{E} \cong \mathcal{E}'$.
Nun ist das Diagramm, vgl. 5.7,

$$\begin{array}{ccccccc}
0 & \longrightarrow & \mathcal{E} & \longrightarrow & 6\mathcal{O}(1) & \xrightarrow{(A_{1,o}Z_1)} & \mathcal{O}(1) \oplus 4\mathcal{O}(2) \\
 & & \Big\downarrow \sigma & & \Big\downarrow \Sigma_2(3) & & \Big\downarrow \begin{pmatrix} \Sigma_o(1) & 0 \\ 0 & \Sigma_1(3) \end{pmatrix} \\
0 & \longrightarrow & \sigma^*\overline{\mathcal{E}} & \longrightarrow & \sigma^*\overline{6\mathcal{O}(1)} & \xrightarrow[(A_{1,o}Z_1)^{\sigma}]{} & \sigma^*\overline{\mathcal{O}(1)} \oplus \sigma^*\overline{4\mathcal{O}(2)}
\end{array}$$

kommutativ, denn $(A_{1,o}, Z_1)^{\sigma} = (\overline{A}_{1,o}, Z_1^{\sigma})$ und es ist
$\Sigma_2(3)\overline{A}_{1,o} = A_{1,o}\Sigma_o(1)$ wegen (iii) und $\Sigma_2(3)Z_1^{\sigma} = Z_1\Sigma_1(3)$
wegen 5.7.
Da jedes Σ_s eine reelle Involution war, sind die hier definierten Involutionen symplektisch und induzieren damit eine symplektische Involution σ für $\mathcal{E}$. Diese stimmt bis auf einen Faktor c, $|c| = 1$, mit der auf $\mathcal{E}$ vorgegebenen symplektischen Involution überein.

(6.2) Holomorphe Trivialisierung von $\mathcal{E}|U_o$.
Nach 3.6 spielt es zur Berechnung des Zusammenhangs zu $(\mathcal{E}, \sigma)$ keine Rolle, welche Trivialisierung von $\mathcal{E}|U_o$ benutzt wird. Hier sei eine kanonische Trivialisierung angegeben, die $a_{o1} \neq 0$ benutzt. Wegen $(i)^*$ ist aber für Instantonenbündel stets $a_{o1}a_{23} \neq 0$. Es gibt nun ein kommutatives Diagramm über U_o mit exakten Zeilen

$$
\begin{array}{ccccccccc}
0 & \longrightarrow & \mathcal{E}|U_o & \xrightarrow{\varepsilon} & 6\mathcal{O} & \xrightarrow{F} & 5\mathcal{O} & & \\
 & & \| & & \downarrow{\scriptstyle \widetilde{pr}} & & & & \\
0 & \longrightarrow & \mathcal{E}|U_o & \xrightarrow{\tilde{\varepsilon}} & 3\mathcal{O} & \xrightarrow{G} & \mathcal{O} & \longrightarrow & 0 \\
 & & \downarrow{\scriptstyle \varphi_o} & & \downarrow{\scriptstyle M} & & \| & & \\
0 & \longrightarrow & 2\mathcal{O} & \xrightarrow[j]{} & 2\mathcal{O}\oplus\mathcal{O} & \xrightarrow[q]{} & \mathcal{O} & \longrightarrow & 0 ,
\end{array}
$$

wobei die letzte Zeile die kanonische spaltende Sequenz ist, ε durch $\mathcal{E}|U_o \longrightarrow 6\,\mathcal{O}(1)\,|U_o \xrightarrow{\approx} 6\mathcal{O}$ vermöge $\frac{1}{z_o}$ definiert ist und $\tilde{\varepsilon} = \widetilde{pr}\circ\varepsilon$ mit

$$\widetilde{pr}(f_{23}\ -f_{13}\ f_{o3}\ f_{12}\ -f_{o2}\ f_{o1}) = (f_{o3}\ f_{o2}\ f_{o1}) .$$

Die Homomorphismen F, G, M sind als Matrizen durch

$$F = (A_{1,o}, Z_1) \qquad G = \begin{pmatrix} g_3 \\ g_2 \\ g_1 \end{pmatrix} \qquad M = \begin{pmatrix} a_{o1} & 0 & g_3 \\ 0 & -a_{o1} & g_2 \\ -a_{o3} & a_{o2} & g_1 \end{pmatrix}$$

definiert mit

$$g_3 = a_{12} - a_{o2}\frac{z_1}{z_o} + a_{o1}\frac{z_2}{z_o}$$

$$g_2 = -a_{13} + a_{o3}\frac{z_1}{z_o} - a_{o1}\frac{z_3}{z_o}$$

$$g_1 = a_{23} - a_{o3}\frac{z_2}{z_o} + a_{o2}\frac{z_3}{z_o} \quad .$$

Die Funktionen g_1, g_2, g_3 werden durch Darstellung von $\mathcal{E}$ vorgeschrieben, wenn man wie hier $\widetilde{pr}$ auszeichnet. Der Leser kann nun verifizieren, daß die obige Ergänzung M der Spalte G eine invertierbare Matrix liefert mit

$$\det M = -a_{o1}(a_{o1}a_{23} - a_{o2}a_{13} + a_{o3}a_{12}) \neq 0 .$$

Weiter kann man leicht verifizieren, daß das Quadrat mit M, G, q kommutativ ist, und daß die mittlere Zeile tatsächlich exakt ist, da Relationen mit G entsprechende Relationen mit F ergeben. Damit wird nun ein Isomorphismus φ_o durch M induziert.

(6.3) Nun wird eine Basis von $F|V_1$ wie folgt konstruiert. Für festes $p \in S^4$ sei

$$\tilde{s}(p) = \frac{1}{\delta(p)} z_o g_3 P_{1,2} - (z_2, -z_1, z_o, 0, 0, 0) ,$$

wobei $P_{1,2} = (p_{23}\ -p_{13}\ p_{o3}\ p_{12}\ -p_{o2}\ p_{o1})$ wie in § 4 gesetzt ist. Dann ist $\tilde{s}(p)$ ein globaler Schnitt in $6\mathcal{O}(1)$.

Es gilt

$$\tilde{s}(p)(A_{1,o}, Z_1) = (0, \frac{1}{\delta(p)} z_o g_3 P_{1,2} Z_1)$$

wie man leicht verifiziert. Da aber $z \in L_p$ gleichbedeutend mit $P_{1,2}Z_1 = 0$ ist, definiert $\tilde{s}(p)$ durch Einschränkung auf L_p einen Schnitt

$$s(p) \in \Gamma(L_p, E|L_p) = F_p$$

Für $p_{o1} = |z_o|^2 + |z_1|^2 \neq 0$ folgt dann $s(p) \neq 0$. Die Zuordnung $p \longrightarrow s(p)$ ist dann ein differenzierbarer Schnitt von F auf $V_1 = S^4 \cap \{p_{o1} \neq 0\}$ ohne Nullstelle. Der Schnitt σs mit $(\sigma s)(p) = \sigma_p s(p)$ hat dann die Form

$$(\sigma s)(p) = \frac{1}{\delta} z_o g_2 P_{1,2} - (-z_3, 0, 0, z_1, -z_o, 0),$$

wie man sofort nach 5.7.2 berechnen kann.

Die Schnitte s, σs von $F|V_1$ sind nun orthogonal bezüglich einer durch σ definierten Metrik, 3.6, und es gilt $\langle\sigma s,\sigma s\rangle = \langle s,s\rangle$. Ist $\rho(p) = \langle s(p), s(p)\rangle$, so sind $s_1 = \frac{s}{\sqrt{\rho}}$ und $s_2 = \frac{\sigma s}{\sqrt{\rho}}$ differenzierbare orthonormale Schnitte in $F|V_1$, die gleichzeitig eine differenzierbare Trivialisierung $E|U_o \cup U_1 \xrightarrow{\psi} (U_o \cup U_1) \times \mathbb{C}^2$ definieren, vgl. 3.6. Mit $h_o = \varphi_o \psi^{-1}$ erhält man dann eine Matrix, aus der man den gewünschten Zusammenhang auf V_1 berechnen kann.

(6.4) Mit den obigen Bezeichnungen gilt:

$$h_o = \frac{1}{\sqrt{\rho}} C = \frac{1}{\sqrt{\rho}} \begin{pmatrix} c_{11} & c_{12} \\ c_{21} & c_{22} \end{pmatrix}$$

mit

$$c_{11} = \frac{1}{\delta} g_3 (p_{o3} a_{o1} - p_{o1} a_{o3}) - a_{o1}$$

$$c_{21} = \frac{1}{\delta} g_3 (p_{o1} a_{o2} - p_{o2} a_{o1})$$

$$c_{12} = \frac{1}{\delta} g_2 (p_{o3} a_{o1} - p_{o1} a_{o3})$$

$$c_{22} = \frac{1}{\delta} g_2 (p_{o1} a_{o2} - p_{o2} a_{o1}) + a_{o1} .$$

Dies erhält man durch direkte Berechnung von $\varphi_o \psi^{-1}$, wobei man auf U_o statt s_1, s_2 die holomorphen Funktionen $z_o^{-1} s_1$, $z_o^{-1} s_2$ zu betrachten hat. Mit den Relationen $P_{1,1} Z_o = 0$ folgt weiter

(6.4.1)
$$\det C = -\frac{a_{o1} p_{o1} d}{\delta} .$$

Nach 3.6 erhält man nun in

$$\tilde{\omega}_1''^{t} = h_o^{-1} \bar{\partial} h_o = \sqrt{\rho}\, C^{-1}\, \bar{\partial}\left(\frac{1}{\sqrt{\rho}} C\right) = -\frac{1}{2} \frac{\bar{\partial}\rho}{\rho} I + C^{-1}\, \bar{\partial}\, C$$

den transponierten (0,1)-Anteil der Zusammenhangsform von $\tilde{\nabla}$ auf $U_o \cup U_1$. Da ω_1 auf V_1 Koeffizienten in $su(2)$ haben muß, erhält man eine Bedingung für die Funktion ρ, die dadurch bis auf einen konstanten Faktor eindeutig bestimmt ist. Setzt man $\rho(p) = \frac{p_{o1}}{\delta(p)}$, so ist diese Bedingung erfüllt. Wir berechnen $\tilde{\omega}_1''^t$ mit dieser Funktion ρ. Dadurch wird es nicht notwendig, einen globalen Schnitt von $\mathcal{E} \wedge \mathcal{E}$ zu berechnen, durch den die Metrik festgelegt wäre. Damit erhält man

(6.4.2) $$\tilde{\omega}_1''^t = \frac{1}{2} \frac{\bar{\partial}\delta^*}{\delta^*} I + C^{-1} \bar{\partial} C,$$

wobei $\delta^* = p_{o1}^{-1}\delta$ gesetzt. Alle Funktionen werden im folgenden vermöge $\pi: \mathbb{P}_3 \to S^4$ als Funktionen auf $U_o \cup U_1$ aufgefaßt, insbesondere $p_{ij} = p_{ij}(z)$ als Plückerkoordinaten zu $z, \sigma z$, so daß stets $P_{1,1} z_o = 0$.

(6.5) Zur Ausrechnung von 6.4.2 benutzen wir weiter folgende Identitäten, wobei

$$z_i^* = \frac{z_i}{z_o}, \quad p_{ij}^* = \frac{p_{ij}}{p_{o1}}, \quad \delta^* = \frac{\delta(p)}{p_{o1}} .$$

(1) $a_{o1}\delta^* - d = g_2(p_{o2}^* a_{o1} - a_{o2}) + g_3(p_{o3}^* a_{o1} - a_{o3})$

(2) $\bar{\partial}\delta^* = g_2 \bar{\partial} p_{o2}^* + g_3 \bar{\partial} p_{o3}^*$

(3) $\bar{\partial} p_{12}^* = z_1^* \bar{\partial} p_{o2}^*$,

$\bar{\partial} p_{13}^* = z_1^* \bar{\partial} p_{o3}^*$

(4) $z_2^{*}\bar{\partial}p_{o3}^{*} = p_{o2}^{*}\bar{\partial}p_{13}^{*} - p_{12}^{*}\bar{\partial}p_{o3}^{*}$

$z_3^{*}\bar{\partial}p_{o2}^{*} = p_{o3}^{*}\bar{\partial}p_{12}^{*} - p_{13}^{*}\bar{\partial}p_{o2}^{*}$

$z_2^{*}\bar{\partial}p_{o2}^{*} = p_{o2}^{*}\bar{\partial}p_{12}^{*} - p_{12}^{*}\bar{\partial}p_{o2}^{*}$

$z_3^{*}\bar{\partial}p_{o3}^{*} = p_{o3}^{*}\bar{\partial}p_{13}^{*} - p_{13}^{*}\bar{\partial}p_{o3}^{*}$.

Diese Identitäten ergeben sich leicht durch Ausrechnung unter Benutzung von $P_{1,1}Z_o = 0$.

(6.6) Seien nun $\eta_{\alpha\beta}$ die Komponenten der Matrix $\tilde{\omega}_1''$, d.h.

$$\tilde{\omega}_1'' = \begin{pmatrix} \eta_{11} & \eta_{12} \\ \eta_{21} & \eta_{22} \end{pmatrix} .$$

Indem man die Koeffizienten $c_{\mu\nu}$ von C in 6.4.2 einsetzt und die Identitäten 6.5 benutzt, folgt durch Ausrechnen:

$$2\delta^{*}\eta_{11} = (p_{13}^{*}a_{o1} - a_{13})\bar{\partial}p_{o2}^{*} - (p_{o3}^{*}a_{o1} - a_{o3})\bar{\partial}p_{12}^{*}$$
$$+(p_{12}^{*}a_{o1} - a_{13})\bar{\partial}p_{o3}^{*} - (p_{o2}^{*}a_{o1} - a_{o2})\bar{\partial}p_{13}^{*}$$

$$\delta^{*}\eta_{12} = (p_{12}^{*}a_{o1} - a_{12})\bar{\partial}p_{o2}^{*} - (p_{o2}^{*}a_{o1} - a_{o2})\bar{\partial}p_{12}^{*}$$

$$\delta^{*}\eta_{21} = (p_{o3}^{*}a_{o1} - a_{o3})\bar{\partial}p_{13}^{*} - (p_{13}^{*}a_{o1} - a_{13})\bar{\partial}p_{o3}^{*}$$

$$\eta_{22} = -\eta_{11} .$$

Die Formenmatrix $\tilde{\omega}_1$ des Zusammenhangs $\tilde{\nabla}$ ist dann gegeben durch

$$\tilde{\omega}_1 = \tilde{\omega}_1' + \tilde{\omega}_1'' = \begin{pmatrix} \eta_{11} - \bar{\eta}_{11}, & \eta_{12} - \bar{\eta}_{21} \\ \eta_{21} - \bar{\eta}_{12}, & \eta_{22} - \bar{\eta}_{22} \end{pmatrix} =: \begin{pmatrix} \theta_{11} & \theta_{12} \\ \theta_{21} & \theta_{22} \end{pmatrix} ,$$

wobei wegen der orthonormalen Basis gilt:

$$\tilde{\omega}_1' = -\bar{\tilde{\omega}}_1''^{t} .$$

Wegen der Bedingung (iii) folgt nun sofort, daß

$$2\delta^*\theta_{11} = (p_{13}^*a_{o1} - a_{13})dp_{o2}^* - (p_{o3}^*a_{o1} - a_{o3})dp_{12}^*$$
$$+(p_{12}^*a_{o1} - a_{12})dp_{o3}^* - (p_{o2}^*a_{o1} - a_{o2})dp_{13}^*$$
$$\delta^*\theta_{12} = (p_{12}^*a_{o1} - a_{12})dp_{o2}^* - (p_{o2}^*a_{o1} - a_{o2})dp_{12}^*$$
$$\theta_{22} = -\theta_{11}, \ \theta_{21} = -\overline{\theta}_{12} \ .$$

Daraus folgt aber weiter sofort, daß $\tilde{\omega}_1$ von der Form $\tilde{\omega}_1 = \pi^*\omega_1$ ist, wobei ω_1 durch die $\theta_{\alpha\beta}$ ohne vorherige Einsetzung von $p = p(z)$ definiert ist, denn p_{o2}^*, p_{12}^*, p_{o3}^*, p_{13}^* sind gerade lokale Koordinaten von V_1, vgl. § 2.

Die Matrix ω_1 hat nach dieser Ausrechnung auch tatsächlich Werte in $\mathfrak{su}(2)$, denn es ist $\omega_1 + \overline{\omega}_1^t = 0$ und $\theta_{11} + \theta_{22} = 0$. Nach den allgemeinen Bemerkungen von § 3 definiert ω_1 auch eine selbstduale Krümmungsform.

Im folgenden sei $a_{o1} = 1$ angenommen. Denn für Instantonenbündel ist stets $a_{o1} \neq 0$ und durch Multiplikation mit a_{o1}^{-1} ändert sich nicht die Äquivalenzklasse des Feldes bzw. des holomorphen Vektorbündels. Durch $a_{o1} = 1$ sind dann die Koeffizienten eindeutig bestimmt und die Formen $\theta_{\alpha\beta}$ haben dann eine symmetrische Gestalt.

(6.7) Will man die Formen $\theta_{\alpha\beta}$ durch die üblichen Koordinaten $t_1,\ldots,t_4$ des $\mathbb{R}^4 \cong V_1 \subset S^4$ ausdrücken, so hat man lediglich nach der Transformation 2.5

$$t_1 - it_2 = p^*_{13} \qquad t_3 - it_4 = p^*_{o3}$$
$$t_1 + it_2 = p^*_{o2} \qquad t_3 + it_4 = -p^*_{12}$$

zu transformieren. Dabei ist es sinnvoll, die Parameter $a_{o2},\dots,a_{23}$ entsprechend durch neue Parameter d, $a_1,\dots,a_4$ mit

$$a_1 - ia_2 = a_{13} \qquad a_3 - ia_4 = a_{o3}$$
$$a_1 + ia_2 = a_{o2} \qquad a_3 + ia_4 = -a_{12}$$
$$d = a_{23} - a_{o2}a_{13} + a_{o3}a_{12} > 0$$

zu ersetzen. Diese sind dann alle reell und unterliegen lediglich der Bedingung $d>0$. Die Formen $\theta_{\alpha\beta}$ und δ^* erhalten dann die Gestalt

$$\theta_{22} = -\theta_{11} \qquad \theta_{21} = -\bar{\theta}_{12}$$
$$\delta^* = d + \|t - a\|^2 = d + (t_1 - a_1)^2 + \dots + (t_4 - a_4)^2$$

$$\theta_{11} = \frac{i}{\delta^*}((t_1 - a_1)dt_2 - (t_2 - a_2)dt_1 + (t_3 - a_3)dt_4 - (t_4 - a_4)dt_3)$$

$$\theta_{12} = \frac{1}{\delta^*}((t_1 - a_1)dt_3 - (t_3 - a_3)dt_1 + (t_4 - a_4)dt_2 - (t_2 - a_2)dt_4)$$
$$+\frac{i}{\delta^*}((t_2 - a_2)dt_3 - (t_3 - a_3)dt_2 + (t_1 - a_1)dt_4 - (t_4 - a_4)dt_1).$$

Ordnet man die Formenmatrix $\omega_1 = (\theta_{\alpha\beta})$ nach den dt_μ , so erhält man schließlich

$$\omega_1 = A_1(t)dt_1 + \dots + A_4(t)dt_4$$

mit

$$A_1(t) = \frac{1}{d+\|t-a\|^2}\left(\begin{array}{c|c} -i(t_2-a_2) & -(t_3-a_3)-i(t_4-a_4) \\ \hline (t_3-a_3)-i(t_4-a_4) & i(t_2-a_2) \end{array}\right)$$

$$A_2(t) = \frac{1}{d+\|t-a\|^2}\left(\begin{array}{c|c} i(t_1-a_1) & (t_4-a_4)-i(t_3-a_3) \\ \hline -(t_4-a_4)-i(t_3-a_3) & -i(t_1-a_1) \end{array}\right)$$

$$A_3(t) = \frac{1}{d+\|t-a\|^2} \begin{pmatrix} -i(t_4-a_4) & (t_1-a_1)+i(t_2-a_2) \\ -(t_1-a_1)+i(t_2-a_2) & i(t_4-a_4) \end{pmatrix}$$

$$A_4(t) = \frac{1}{d+\|t-a\|^2} \begin{pmatrix} i(t_3-a_3) & -(t_2-a_2)+i(t_1-a_1) \\ (t_2-a_2)+i(t_1-a_1) & -i(t_3-a_3) \end{pmatrix}$$

Die hier angegebenen Lösungen stellen nach dem Darstellungssatz für holomorphe Vektorbündel und dem Satz von Atiyah-Ward alle selbst-dualen Lösungen der euklidischen Yang-Mills Gleichung zur Instantonenzahl $c_2 = 1$ dar. Es sind gerade die t'Hooft Lösungen. Verschiedene Werte der Parameter führen zu nicht eichäquivalenten Feldern.

Bemerkung: Die Koeffizienten $\theta_{\alpha\beta}$ der Formenmatrix ω_1 sind nach der Konvention 1.3. indiziert. In der Literatur ist es oft üblich, ω_1 als transponierte Matrix zu definieren. Dies ist zu beachten, wenn man die hier hergeleiteten Matrizen mit anderen Darstellungen vergleicht.

L i t e r a t u r

[1] Atiyah, M. F. - Vorträge in Bonn, Oxford

[2] Atiyah, M. F., Hitchin, N. J., Singer, I. M. - Selfduality in four-dimensional Riemannian geometry. To appear in Proc. Roy. Soc.

[3] Atiyah, M. F., Hitchin, N. J., Drinfeld, V. G., Manin, Y. I. - Construction of Instantons, Phys. Letters, 65 A, 185 - 187 (1978)

[4] Atiyah, M. F., Ward, R. S. - Instantons and Algebraic Geometry, Commun. Math. Phys. 55, 117 - 124 (1977)

[5] Barth, W. - Moduli of vector bundles on the projective plane, Invent. - math. 42, 63 - 91 (1977)

[6] Barth, W. - Some properties of stable rank-2 vector bundles on $\mathbb{P}_n$, Math. Ann. 226, 125 - 150 (1977)

[7] Barth, W., Hulek, K. - Monads and moduli of vector bundles, manuscripta math. 25, 323 - 347 (1978)

[8] Douady, A. - Vorträge in Nizza, Kaiserslautern

[9] Grauert, H., Mülich, G. - Vektorbündel vom Rang 2 über dem n-dimensionalen komplex-projektiven Raum, manuscripta math. 16, 75 - 100 (1975)

[10] Hartshorne, R. - Stable vector bundles of rank-2 on $\mathbb{P}_3$, Math. Ann. 238, 229 - 280 (1978)

[11] Hartshorne, R. - Stable vector bundles and Instantons, Commun. math. Phys. 59, 1 - 15 (1978)

[12] Hirzebruch, F. - Topological methods in algebraic geometry, Springer 1966

[13] Horrocks, G. - Examples of rank three vector bundles on five-dimensional projective space, J. London Math. Soc. (2), 18, 15 - 27 (1978)

[14] Kobayashi, S., Nomizu, K. - Foundations of differential geometry, I, II, Interscience 1963/69

[15] Kohn, J. J., Nirenberg, L. - On the algebra of pseudodifferential operators, Comm. Pure Appl. Math. 18, 269 - 305 (1965)

[16] Maruyama, M. - Moduli of stable sheaves I and II, Journ. Math. Kyoto Univ., 17, 91 - 126 (1977) and 18, 557 - 614 (1978)

[17] Milnor, J. W., Stasheff, J. D. - Characteristic classes, Princeton University Press 1974

[18] Newlander, A., Nirenberg L. - Complex coordinates in almost complex manifolds, Ann. Math. 65, 391 - 404 (1957)

[19] Rawnsley, J. H. - Self-dual Yang-Mills fields, Manuskript

[20] Schneider, M. - Holomorphic vector bundles on $\mathbb{P}_n$, Sem. Bourbaki, n° 530, 1978/79

[21] Steenrod, N. - The topology of fibre bundles, Princeton 1951

[22] Trautmann, G. - Moduli for vector bundles on $\mathbb{P}_n(\mathbb{C})$, Math. Ann. 237, 167 - 186 (1978)

[23] Trautmann, G. - Moduli von Vektorbündeln auf $\mathbb{P}_n$ und Parametrisierung von Maruyama-Schemata, in Vorbereitung

[24] Wells, R. O. - Differential analysis on complex manifolds, Prentice-Hall 1973

[25] Uhlenbeck, K. K. - Removable singularities in Yang-Mills fields, Bull. Amer. Math. Soc. (new series) 1, 579 - 581 (1979)

Anschrift des Autors:

Günther Trautmann
Fachbereich Mathematik
Universität Kaiserslautern
6750 Kaiserslautern
BR Deutschland

Manuscripts

Manuscripts should be no less than 100 and preferably no more than 500 pages in length.

They are reproduced by a photographic process and therefore must be typed with extreme care. Symbols not on the typewriter should be inserted by hand in indelible black ink. Corrections to the typescript should be made by pasting in the new text or painting out errors with white correction fluid.

The typescript is reduced slightly (75%) in size during reproduction; best results will not be obtained unless the text on any one page is kept within the overall limit of 6¼ x9½ in (16 x 24 cm). On request, the publisher will supply special paper with the typing area outlined.

Manuscripts should be sent to the editors or directly to:
Birkhauser Boston Inc., 380 Green Street, Cambridge, MA 02139